THE MATHEMATICAL THEORY OF DIFFUSION AND REACTION IN PERMEABLE CATALYSTS

THE MATHEMATICAL THEORY OF DIFFUSION AND REACTION IN PERMEABLE CATALYSTS

VOLUME II

QUESTIONS OF UNIQUENESS, STABILITY, AND TRANSIENT BEHAVIOUR

BY

RUTHERFORD ARIS

UNIVERSITY OF MINNESOTA

CLARENDON PRESS · OXFORD

1975

Oxford University Press, Ely House, London W.1

GLASGOW NEW YORK TORONTO MELBOURNE WELLINGTON
CAPE TOWN IBADAN NAIROBI DAR ES SALAAM LUSAKA ADDIS ABABA
DELHI BOMBAY CALCUTTA MADRAS KARACHI LAHORE DACCA
KUALA LUMPUR SINGAPORE HONG KONG TOKYO

ISBN I 19 851942 7

PRINTED IN GREAT BRITAIN BY
J. W. ARROWSMITH LTD., BRISTOL, ENGLAND

TO CLAIRE

PREFACE

THE intention of this monograph is threefold. Its central objective is to bring together in a convenient form the many results on the theory of diffusion and reaction that are now scattered through the literature of chemical engineering, applied mathematics and biophysics. The flanking purposes are to draw to the attention of the chemical engineer the wide variety of mathematical methods that has been used to explore the subject, and to discover to the mathematician a domain of natural philosophy in which some of the latest and most vital results of the theory of partial differential equations may find concrete realization.

At the expense of being a shade sesquipedalian the title is an accurate account of the scope of the book. It is the theory, rather than the practice, of catalysis and the mathematical, rather than physical or chemical, theory that will be discussed. This is appropriate, for the practical aspects of the matter have been ably treated by Satterfield in his *Mass transfer in heterogeneous catalysis*, but it does not mean that the theory is divorced from practice, or that the mathematical models are not based on the physics and chemistry of the situation. The terms 'diffusion and reaction in permeable catalysts' have been used to exclude in a general way the subjects of gas adsorption and gas–solid non-catalytic reaction. Some of the problems treated have a bearing on these topics, as the bibliography will show, but they have not been discussed in a systematic manner. It is not that they are without mathematical interest or practical importance, but merely a question of keeping this monograph within bounds. Finally the term 'permeable', rather than 'porous', has been used intentionally in the hope that it will strike a sympathetic chord in the biologist's ear. It is becoming increasingly obvious that the diffusion and reaction problems of biology have much in common with those of chemical engineering. This awareness of the pervasive influence of diffusion in biology, as witnessed, for example, by Crick's influential letter to *Nature* (1970), is seen everywhere in the literature and there is some danger of unnecessary duplication of effort. I cannot pretend that I have covered the biological aspects in the detail that they deserve, for many of them merit monographs in their own right, but I hope the frequent references to problems with biological overtones will encourage the mutual interest of biologist and engineer in each other's approaches.

The principal objective of bringing together the theoretical results is justified by the importance of the subject and the state of the art. Satterfield's book is 'a carefully documented exposition of what is known, correlated and useful in the domain of catalytic reaction and diffusion', as Carberry said in a review, and the Thomas's *Introduction to the principles of heterogeneous catalysis* admirably fulfils the promise of its title both from the chemical and engineering points of view. But these, and other, books scarcely touch on the mathematical theory, and of those that do Frank–Kamenetskii's classic, *Diffusion and heat exchange in chemical kinetics*, is broad in scope but published before much recent spate of work of importance, while Gavalas' and Villadsen's excellent monographs are deliberately limited to rather special topics. At this stage of development of the subject there is, therefore, a need to give as comprehensive an account as possible of the territory that has been fairly thoroughly mapped, and to offer some indication of where the avenues of further exploration may lie. The mass of work that has been done on steady state solutions of diffusion and reaction problems lies in the former area and the first difficulty I encountered was the realization of how much had been done at various times in various places that required mention in the five chapters which the first volume comprises. But if the pedestrian part of the subject is 'copious without order', as Dr. Johnson found the English language, the burden of the second volume—the questions of uniqueness, stability and the behaviour of transients—is indeed 'energetick without rules'. For here one faces the difficulty of giving an account of matters that are appearing month by month in the scientific journals. There can be no hope of being comprehensively systematic, but it is useful to gather the results together and to attempt to present them in some sort of perspective. How far I have succeeded in this will be for the reader to judge, but I would have had no chance of success without the kindness of numerous correspondents who have sent me manuscripts of their papers and given me permission to refer to them. By their help I have been able to incorporate results that have only appeared in print whilst this book was itself in production.

One serious defect of this work is painfully obvious even to the author, namely the incompleteness of his reference to the Russian literature. One stream of research that has poured into the subject has its source in the work of Zeldovich and Frank-Kamenetskii, not to mention the prophetic insight of Lomonosov, who was convinced by his own art that 'chemical experiments combined with physical show particular effects'. Already in 1947 Semenov, in the preface to the first edition of Frank-Kamenetskii's book, was able to claim that Soviet science has assumed 'an honourable place in the field of chemical kinetics in general and macroscopic kinetics in particular' and his hope that the book would 'stimulate the further development of this field' has been amply fulfilled in work of Levich, Boreskov, Slinko, Pismen and Kharkats to mention but a few. I have a number of references to Russian work, but I am sure that they fail to do justice to its scope and perhaps do violence to its priorities. If this book is translated into Russian, the translation editors will do a great service to the subject if they will insert or append a more complete bibliography of the Russian literature.

I have incurred so many and various obligations to other workers in this field that I must begin by saying that I hope any whom I have neglected will forgive the oversight. I owe a variety of things, ranging from figures to footnotes and from general encouragement to specific correction to these colleagues, and though detailed references are given at appropriate points, I would like here to mention their names: N. R. Amundson, D. G. Aronson, J. E. Bailey, K. B. Bischoff, L. F. Brown, J. B. Butt, J. J. Carberry, D. S. Cohen, D. L. Cresswell, M. M. Denn, J. A. DeSimone, D. W. Drott, G. R. Gavalas, C. Georgakis, V. Hlaváček, R. Jackson, D. D. Joseph, J. P. G. Kehoe, H. B. Keller, K. H. Keller, T. W. Laetsch, D. Luss, C. McGreavy, M. Marek, B. N. Mehta, M. Mercer, F. Mussatti, H. G. Othmer, J. R. A. Pearson, L. Peletier, E. E. Petersen, A. B. Poore, S. Rester, G. W. Roberts, C. N. Satterfield, L. E. Scriven, J. M. Smith, W. E. Stewart, J. C. R. Turner, A. Varma, J. Villadsen, E. Wicke. Professor Wicke most kindly put at my disposal his vast knowledge of the subject and particularly of its history; D. Luss has kept me abreast of his own work and been more than helpful in his criticism, and in supplying missing information and allowing me to use his figures; he and J. E. Bailey have read much of the manuscript and I am most grateful to them for their comments; C. N. Satterfield, to whose book I have constantly turned, has put me further in his debt by sending me some most apposite and helpful remarks. I need not add that, though many blemishes have been removed by the help of these colleagues, I am of course responsible for those that remain. To Professor Danckwerts I owe the hospitality of the Shell Department of Chemical Engineering where the main part of the writing was done under circumstances such that the product of my effort must inevitably fall short of the opportunity. I am grateful to the Regents of the University of Minnesota for a sabbatical leave and to the Trustees of the Guggenheim Foundation for their generous support during this time. The Press and I are further indebted to them for a direct subvention which has kept the price of the book within bounds. I am also obliged to the National Science Foundation for their long-term support of my efforts to understand the behaviour of chemical reactors; some of the results reported here were obtained during this research.

The typing has largely been the work of Mrs. J. Grose, Miss Linda Anderson and Miss Gail Witchurch though some early fragments were typed in Cambridge by Miss Margaret Sansome and Miss Philippa McNair. In the many revisions and rearrangements Ms. Flurnia Davis and Miss Marsha Riebe have also been of the greatest assistance. The staff of the Clarendon Press has been most patient and helpful in the whole process of production and the book benefits greatly from their editorial skill.

My obligations to my wife are too personal to be spelled out here: suffice it to say that she has survived another book and is forgiven her interruptions even as she has pardoned my preoccupation and peevishness. It is to her that I dedicate it in gratitude and affection.

R.A.

CONTENTS

VOLUME I

VOLUME II

GENERAL NOMENCLATURE

THE following list gives the principal usage of a number of symbols that are common to all chapters of the book. This saves their being repeated in the particular nomenclature of each chapter. It has not always been possible to avoid using the same letter for two purposes even within the same chapter, but then the usages are clearly stated in the list at the end of that chapter. Purely ephemeral notation, such as a variable of integration, is not recorded.

A_i or A_j	ith or jth chemical species
a	characteristic dimension of catalytic body
c	concentration (used with suffixes q.v.)
$\hat{c}$	surface concentration
D	diffusion coefficient (used with suffixes q.v.)
$\hat{D}$	surface diffusion coefficient
D_e	effective diffusion coefficient
E	activation energy
J, j or $\mathbf{j}$	flux vector or magnitude of flux
k	rate constant of reaction or adsorption (used with suffixes q.v.)
Le	Lewis number
$M(. ;.)$	integral in the solution of the slab problem and normalization constant for Thiele modulus
Nu	Nusselt number
p	order of reaction
Pr	Prandtl number
q	index of symmetrical shapes (0, slab; 1 cylinder; 2 sphere)
$R(u)$, $R(u, v)$	dimensionless reaction rate
R	gas constant
Re	Reynolds number
$\mathbf{r}$	position vector within catalytic body
r	magnitude of $\mathbf{r}$
$r(c, T)$, $\hat{r}(c, T)$	reaction rate, rate per unit area
Sc	Schmidt number
S_g	specific area of catalyst
S_x	external surface area of catalytic body ($=\sigma a^2$)
T	temperature (used with suffixes q.v.)
t	time
u	dimensionless concentration (used with suffixes q.v.)
V_p	volume of catalytic body ($=\upsilon a^3$)
v	dimensionless temperature (used with suffixes q.v.)
α_i, α_{ji}	stoicheiometric coefficient

β	Prater number, $(-\Delta H)D_e c_f/k_c T_f$
γ	Arrhenius number, E/RT_f
Δ	dimensionless diffusivity
ΔH	heat of reaction
ΔH_a	heat of adsorption
η	effectiveness factor
μ	Biot number for heat transfer
ν	Biot number for mass transfer
$\boldsymbol{\rho}, \rho$	dimensionless space variable within catalytic body
ρ_b	bulk density of catalytic body
σ	dimensionless surface area, S_q/a^2
υ	dimensionless volume, V_p/a^3
ϕ	Thiele modulus
Φ	Thiele modulus normalized so that $\eta\Phi \sim 1$
$\Omega, \partial\Omega$	region occupied by catalytic body, its boundary
ω	fractional radius of dead zone
$\partial/\partial n$	differentiation normal to $\partial\Omega$
dS	surface element
dV	volume element
$\partial/\partial \nu$	differentiation normal to $\partial\Omega$ in dimensionless variables
$d\Sigma$	surface element in dimensionless variables
$d\Upsilon$	volume element in dimensionless variables

SUFFIXES AND AFFIXES

e	effective or equilibrium value
f	value far from body
i,j	pertaining to species or reaction, i or j
o	centre value
s	surface value
ˆ	value per unit surface area

6

EXISTENCE AND UNIQUENESS OF THE STEADY STATE

6.1. Data and desiderata

IN the three chapters of this second volume we enter an area which is still very much the subject of current research, and while there may be details to be developed in the topics covered up to this point, the whole paradigm of what lies before us has yet to be set in order. It is futile to essay any sort of comprehensiveness and it is rather a matter of seeking perspective. Let us first review some of the phenomena that have emerged which raise questions of uniqueness, stability, and transient behaviour, for these three subjects are linked together. Uniqueness and stability are properties of the steady state solution and while it is sometimes the case that a unique steady state is stable this cannot be taken for granted, but must be proved. On the other hand when there are more steady states than one, some must be unstable. The notion of stability, however, is essentially bound up with the transient behaviour of the system, for it claims that deviations from the steady state tend to zero or at least remain bounded. If a unique steady state is unstable a stable limit cycle is very often present and this can only be discovered by some understanding of the transient behaviour.

We have seen that if the domain Ω is suitable and the kinetic function $R(u, v)$ sufficiently nonlinear, then the Dirichlet problem

$$\nabla^2 u = \phi^2 R(u, v), \qquad \nabla^2 v + \beta\phi^2 R(u, v) = 0 \quad \text{in } \Omega, \tag{6.1}$$

$$u = v = 1 \quad \text{on } \partial\Omega \tag{6.2}$$

may have more than one solution in some interval (ϕ_b, ϕ^b). Indeed when Ω is a sphere and R is exponential there can be infinitely many solutions for a certain value of ϕ and, though $\exp u$ is not a realistic function in a chemical context, there are rate expressions soundly grounded in physical chemistry that exhibit an arbitrarily large number of solutions by suitable adjustment of parameters (Section 4.3.4). In the case of the Robin problem, where eqns (6.2) are replaced by

$$\frac{\partial u}{\partial \nu} = \nu(1-u), \qquad \frac{\partial v}{\partial \nu} = \mu(1-v) \quad \text{on } \partial\Omega, \tag{6.3}$$

there can be two intervals in the range of ϕ in which there are multiple solutions, and these may overlap to produce greater multiplicity.

These phenomena raise several questions. First, what conditions on the nonlinear function $R(u, v)$ are sufficient to ensure the uniqueness of the solution for all values of ϕ? Second, if $R(u, v)$ transgresses these conditions, how can the number of solutions be most easily or most exactly determined? Third, which steady states are stable? The last question takes us into the next chapter, but the first two are squarely in this. In particular we would like to answer the second one by showing how estimates of the interval (ϕ_b, ϕ^b) may be made. The answer to the first will be to delimit the nonlinearity of the function, and when a particular form of expression is chosen this will imply the prescription of some domain within its parameter space.

It may be remarked in passing that the possibility of multiple steady states in the catalyst particle implies an even greater multitude of possibilities for the steady state of a reactor packed with many particles. There have been studies of reactor behaviour from this point of view but we shall not be concerned with the reactor as a whole. There is however a considerable similarity of behaviour in our problem to that of a tubular reactor, and a number of papers apply the same methods to get similar results. Some references are given in the bibliography.

Before giving some precise data for the exothermic reactor it is worth introducing the stirred tank reactor analogy. If a first-order irreversible reaction takes place in an adiabatic stirred tank of volume V, the rate at which the reactant disappears is $VA \exp(-E/RT)c$, where c is the concentration and T the temperature in the tank, and the rate constant has been written $A \exp(-E/RT)$. Thus if c_f and T_f are the feed concentration and temperature, and its flow rate is q, the usual balances give the equations

$$V\frac{\mathrm{d}c}{\mathrm{d}t} = q(c_f - c) - VA \exp(-E/RT)c, \tag{6.4}$$

$$VC_p\frac{\mathrm{d}T}{\mathrm{d}t} = qC_p(T_f - T) + (-\Delta H)VA \exp(-E/RT)c, \tag{6.5}$$

where C_p is the heat capacity per unit volume and ΔH the heat of reaction. With the dimensionless variables

$$\tau' = qt/V, \qquad u = c/c_f, \qquad v = T/T_f, \tag{6.6}$$

and parameters

$$\psi^2 = VA \exp(-E/RT_f)/q, \qquad \beta = (-\Delta H)c_f/C_pT_f, \qquad \gamma = E/RT_f, \tag{6.7}$$

we have the ordinary differential equations

$$\frac{\mathrm{d}u}{\mathrm{d}\tau'} = 1 - u - \psi^2 R(u, v), \tag{6.8}$$

$$\frac{\mathrm{d}v}{\mathrm{d}\tau'} = 1 - v + \beta\psi^2 R(u, v), \tag{6.9}$$

$$R(u, v) = u \exp[\gamma(1 - 1/v)]. \tag{6.10}$$

These ordinary differential equations may be compared with transient equations of which eqns (6.1) are the steady state form, namely

$$\frac{\partial u}{\partial \tau} = \nabla^2 u - \phi^2 R(u, v), \tag{6.11}$$

$$\frac{\partial v}{\partial \tau} = \nabla^2 v + \beta\phi^2 R(u, v). \tag{6.12}$$

If we consider the Dirichlet problem with $u = v = 1$ on $\partial\Omega$ for these equations, we may ask what light is thrown on them by the 'linearized' equations obtained by merely ignoring the reaction rate terms. They would then become

$$\frac{\partial \tilde{u}}{\partial \tau} = \nabla^2 \tilde{u} \quad \text{and} \quad \frac{\partial \tilde{v}}{\partial \tau} = \nabla^2 \tilde{v}$$

and their solutions would be

$$\begin{aligned} \tilde{u} &= 1 + \sum_{n=1}^{\infty} a_n \exp(-\lambda_n^2 \tau) w_n(\boldsymbol{\rho}), \\ \tilde{v} &= 1 + \sum_{n=1}^{\infty} b_n \exp(-\lambda_n^2 \tau) w_n(\boldsymbol{\rho}), \end{aligned} \tag{6.13}$$

where λ_n^2 are the eigenvalues and $w_n(\boldsymbol{\rho})$ the eigenfunctions of the problem

$$\nabla^2 w + \lambda^2 w = 0 \quad \text{in } \Omega, \qquad w = 0 \quad \text{on } \partial\Omega. \tag{6.14}$$

Continuing in this rash vein we observe that after a brief initial period the first term, corresponding to the lowest eigenvalue λ_1^2, will dominate the series and then

$$\begin{aligned} \nabla^2 \tilde{u} &= -\lambda_1^2 a_1 \exp(-\lambda_1^2 \tau) w_1(\boldsymbol{\rho}) = \lambda_1^2 (1 - \tilde{u}) \\ \text{and} \quad \nabla^2 \tilde{v} &= -\lambda_1^2 b_1 \exp(-\lambda_1^2 \tau) w_1(\boldsymbol{\rho}) = \lambda_1^2 (1 - \tilde{v}). \end{aligned} \tag{6.15}$$

Substituting these approximations for the Laplacians in eqns (6.11) and (6.12) gives the pair of ordinary differential equations

$$\frac{\mathrm{d}\tilde{u}}{\mathrm{d}\tau} = \lambda_1^2 (1 - \tilde{u}) - \phi^2 R(\tilde{u}, \tilde{v}), \tag{6.16}$$

$$\frac{\mathrm{d}\tilde{v}}{\mathrm{d}\tau} = \lambda_1^2 (1 - \tilde{v}) + \beta\phi^2 R(\tilde{u}, \tilde{v}). \tag{6.17}$$

These equations are precisely those of the stirred tank with

$$\tau' = \lambda_1^2 \tau, \qquad \psi = \phi/\lambda_1. \tag{6.18}$$

Recalling that the definition of ϕ^2 for a first-order reaction would be $a^2 A \exp(-E/RT_f)/D$, we see that the equivalent stirred tank is one of holding time $(V/q) = a^2/D\lambda_1^2$. Such an unbridled derivation as this is chiefly notable for its total abandonment of rigour, but it does show a connection between the stirred tank and catalyst particle which corresponds to one's physical intuition. It was first adumbrated by Frank-Kamenetskii and has been used extensively by Hlaváček and his coworkers. It will appear in different forms arising from various approximation and lumping techniques and is clearly cognate with modal analysis.

Now the steady state eqns (6.8) and (6.9) give

$$1-u-\psi^2 R(u, v) = 0, \tag{6.19}$$

$$1-v+\beta\psi^2 R(u, v) = 0, \tag{6.20}$$

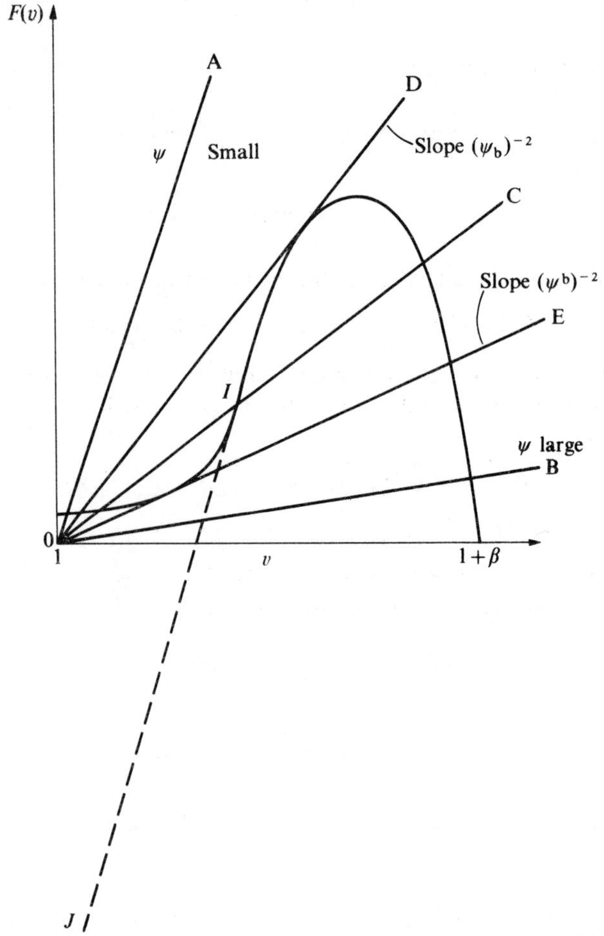

FIG. 6.1. Graphical solution for the steady states of a stirred tank.

and these imply that

$$\beta u + v = 1 + \beta. \tag{6.21}$$

Then using the first order expression given in eqn (6.10) we can eliminate u and write eqn (6.20) as

$$\psi^{-2}(v-1) = F(v) = (1+\beta-v)\exp[\gamma(1-1/v)]. \tag{6.22}$$

The properties of $F(v) = F(v;\beta,\gamma)$ were fully discussed in Section 2.5.4 and the forms which it can take are shown in Fig. 2.2. The steady states of the stirred tank reactor are evidently given by the intersections of the line of slope ψ^{-2} through the point $v = 1$ with the curve $F(v)$. This is shown in Fig. 6.1 from which it is clear that, if ψ is sufficiently small (giving a line such as OA) or sufficiently large (OB), there can only be one intersection and hence a unique steady state. However if the curve of $F(v)$ is as shown, a line such as OC will intersect it three times and a multiplicity of steady states obtains. A necessary and sufficient condition for this to happen is that the tangent at the point of inflexion (IJ) should intersect the vertical line $v = 1$ below O. In fact the slopes of the tangent lines OD and OE give the range (ψ_b, ψ^b) within which ψ must lie for there to be three steady states. But from eqn (2.165) we know that IJ passes above O if

$$\beta\gamma < 4(1+\beta) \tag{6.23}$$

and hence this is a necessary and sufficient condition for uniqueness.

But this behaviour is entirely analogous to what we have found for the catalyst particle, namely, that if $\beta\gamma$ is sufficiently large compared with β, there can be three (or more) steady states provided ϕ lies in the range (ϕ_b, ϕ^b). In fact if the results of calculations for the catalyst slab and the stirred tank were superimposed in such a figure as Fig. 6.2 we should have two bifoliate

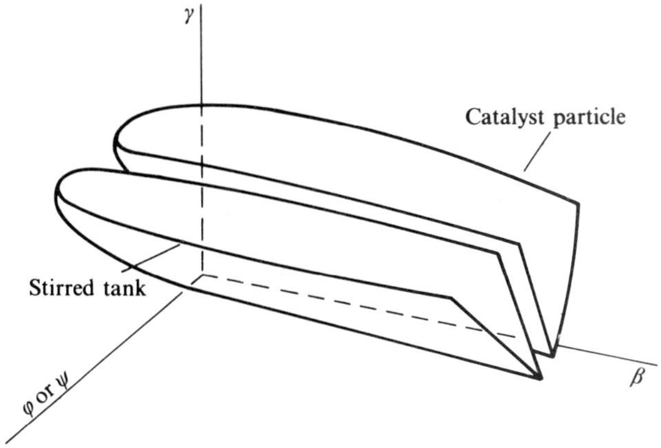

FIG. 6.2. Regions of uniqueness for the stirred tank and the catalyst particle.

surfaces. Outside the two folia the values of ϕ or ψ and β, γ are such that there is only one steady state, but if the point (ϕ, β, γ) or (ψ, β, γ) falls between the folia there are three steady states. Moreover, though the transformation $\psi = \phi/\lambda_1$ would not bring these two surfaces into coincidence, it would bring them quite close together. This analogy with the stirred tank which also arises from considering one-point collocation methods (cf. Section 4.1.2) will be used again later in the discussion of uniqueness and stability.

6.1.1. *Data on the exothermic reaction*

It will be well to summarize the computations that have been done for the exothermic reaction and mention the ways in which they have been presented. The bifoliate surface in Fig. 6.2 can be more easily represented in the space of β, δ, and ϕ or Φ, where $\delta = \beta\gamma$ and Φ is the normalized Thiele modulus, $\phi/M(0; 1)$; cf. eqn (4.99). In Fig. 6.3 the general form of this bifurcation surface is shown for the first-order exothermic reaction in a slab. The line parallel to the Φ-axis through a particular pair of values (β, δ) intersects

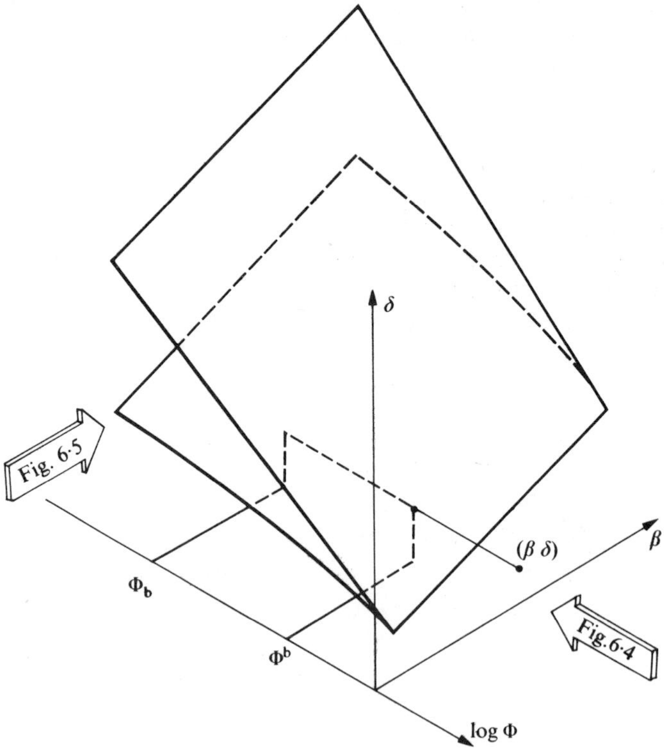

FIG. 6.3. Regions of uniqueness and multiplicity for the first-order exothermic reaction in a slab in the (β, δ, Φ)-space.

the two folia in Φ_b and Φ^b and there are three steady states in the range $\Phi_b < \Phi < \Phi^b$. These values of Φ_b and Φ^b are called bifurcation values since the number of solutions changes as Φ passes through such a value. The arrows in the figure show the viewpoints of the sections in the next two figures and the advantage of using the parameter δ is that these sections are almost straight. If the surface is viewed along the Φ axis its two leaves lie wholly above a leading edge shown as the broken line in Fig. 6.4. This is the locus of parameter pairs such that the (η, Φ)-curve has a vertical tangent at its inflection point. Thus η is a unique function of Φ but the situation is critical in the sense that a slight change in the values of β and γ could lead to a multiplicity of solutions. For such points (β, δ, Φ) the rather ugly name

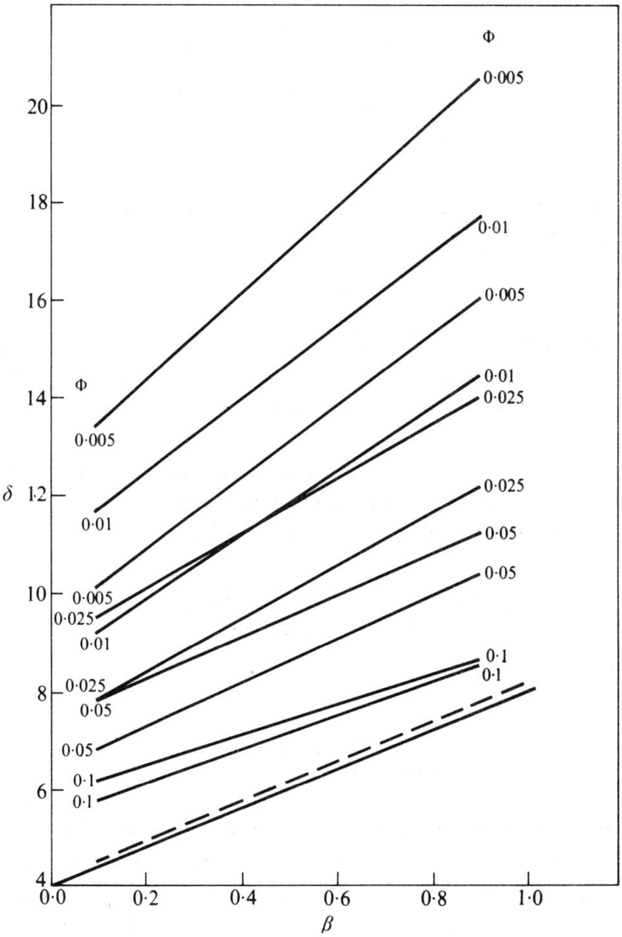

FIG. 6.4. Sections of the bifurcation surface by planes of constant Φ.

trifurcation has been used apologetically by Drott and Aris (1969, p. 544), since they represent the confluence of two bifurcation points. The exact locus of this edge is given in Table 6.1.

TABLE 6.1
Parameters for a point of inflection with vertical tangent

β	γ	δ	Φ
0·1	45·39	4·539	0·198
0·2	24·74	4·947	0·184
0·3	17·85	5·355	0·172
0·4	14·40	5·762	0·161
0·5	12·34	6·169	0·152
0·6	10·96	6·577	0·144
0·7	9·98	6·984	0·137
0·8	9·24	7·391	0·131
0·9	8·67	7·799	0·125

The projection of this curve on the (β, δ)-plane (the broken line in Fig. 6.4) is very close to the straight line

$$\delta = 4{\cdot}074\beta+4{\cdot}132, \tag{6.24}$$

and for (β, δ) below this line the solution is unique for all Φ. We notice at once that this is extraordinarily close to the condition (6.23) for the stirred tank to have a unique steady state; the lowest line in Fig. 6.4 is this condition, $\delta = 4(1+\beta)$.

For the geometry of the slab which permits the solution of the problem by integrals, as in eqns (4.105) and (4.106), the bifurcation values Φ_b and Φ^b can be found from a third integral. For, if by eqn (4.107)

$$\Phi = \frac{M(1;u_0)}{M(1;0)}+\int_{u_0}^{1}\frac{M(u;u_0)}{M(1;0)}N(u)\,\mathrm{d}u,$$

then $\mathrm{d}\Phi/\mathrm{d}u_0 = 0$ when

$$\int_{u_0}^{1}\frac{M(1;u_0)}{M(u;u_0)}N(u)\,\mathrm{d}u = -1. \tag{6.25}$$

But u_0 can be regarded as the parameter along the (Φ, η)-curve, which will have a vertical tangent and hence a bifurcation point when $\mathrm{d}\Phi/\mathrm{d}u_0 = 0$. In using the quadrature formulae (4.105) and (4.106) it is easy enough to keep track of the integral (6.25) and hence to calculate the bifurcation values with considerable accuracy. The values calculated by Drott are given in Table 6.2.

TABLE 6.2
Bifurcation values for first-order reaction in a slab

β	γ	δ	Φ_b	Φ^b
0·1	50	5	0·15443	0·16076
0·1	70	7	0·04461	0·06972
0·1	90	9	0·01124	0·03066
0·1	110	11	0·00268	0·01339
0·2	30	6	0·10614	0·11957
0·2	40	8	0·03242	0·05546
0·2	50	10	0·00891	0·02590
0·2	60	12	0·00232	0·01203
0·3	18·33	5·5	0·16106	0·16199
0·3	23·33	7·0	0·07551	0·09250
0·3	30·00	9·0	0·02447	0·04536
0·3	36·67	11·0	0·00734	0·02233
0·5	12·34	6·17	0·15237	0·15237
0·5	13·00	6·50	0·13333	0·13547
0·5	14·00	7·00	0·10753	0·11441
0·5	15·00	7·50	0·08588	0·09712
0·5	16·00	8·00	0·06807	0·08269
0·5	18·00	9·00	0·04203	0·06025
0·5	20·00	10·00	0·02547	0·04406
0·5	25·00	12·50	0·00687	0·02024
0·5	30·00	15·00	0·00174	0·00929
0·5	35·00	17·50	0·00042	0·00425
0·7	10·00	7·00	0·13645	0·13647
0·7	14·40	10·08	0·03903	0·05513
0·7	20·00	14·00	0·00643	0·01850
0·7	30·00	21·00	0·00020	0·00265
0·9	11·11	10	0·05712	0·06860
0·9	16·67	15	0·00749	0·01932
0·9	22·22	20	0·00084	0·00554
0·9	40·00	36	0·00000047	0·0000988

TABLE 6.3
Bifurcation values for a first-order reaction in a sphere

β	γ	δ	ϕ_b	ϕ^b
0·4	20	8	0·531	0·742
0·6	20	12	0·240	0·587
0·8	20	16	0·125	0·501
0·3	30	9	0·335	0·675
0·4	30	12	0·161	0·573
0·6	30	18	0·046	0·460
0·2	40	8	0·386	0·716
0·3	40	12	0·125	0·568

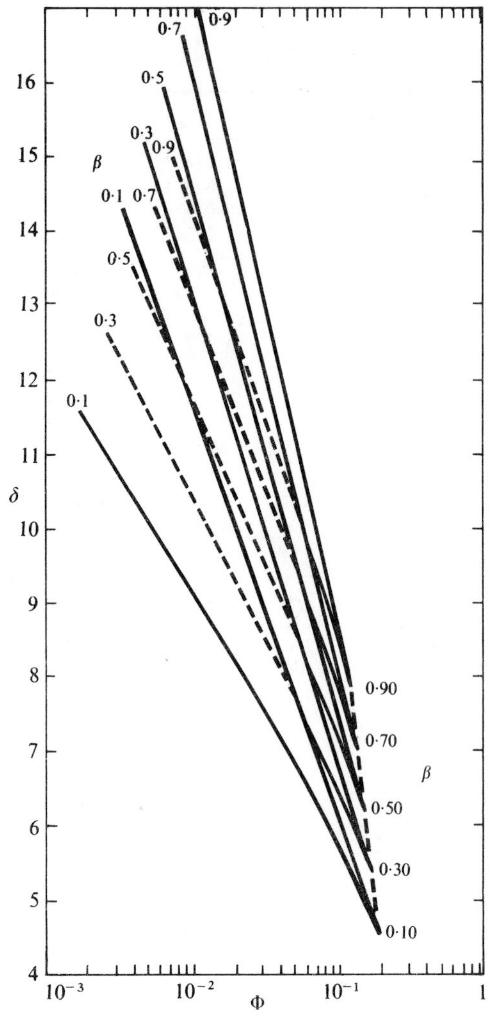

FIG. 6.5. Sections of the bifurcation surface by planes of constant β.

Hlaváček, Marek, and Kubíček (1968*b*) give the values in Table 6.3 for first-order reaction in a sphere using the Thiele modulus $\phi^2 = a^2 A \exp(-E/RT_f)/D_e$, a being the radius of the sphere.

These authors have given results on the trifucation point for pth-order reaction in the symmetric shapes and for the first order in a special case of the Robin problem, namely that with $\mu = \nu$. These are best summarized in the form of Fig. 6.6 and the remark that the critical value of δ is remarkably insensitive to ν in the range $(5, \infty)$. In Fig. 6.6 the critical value of δ is given

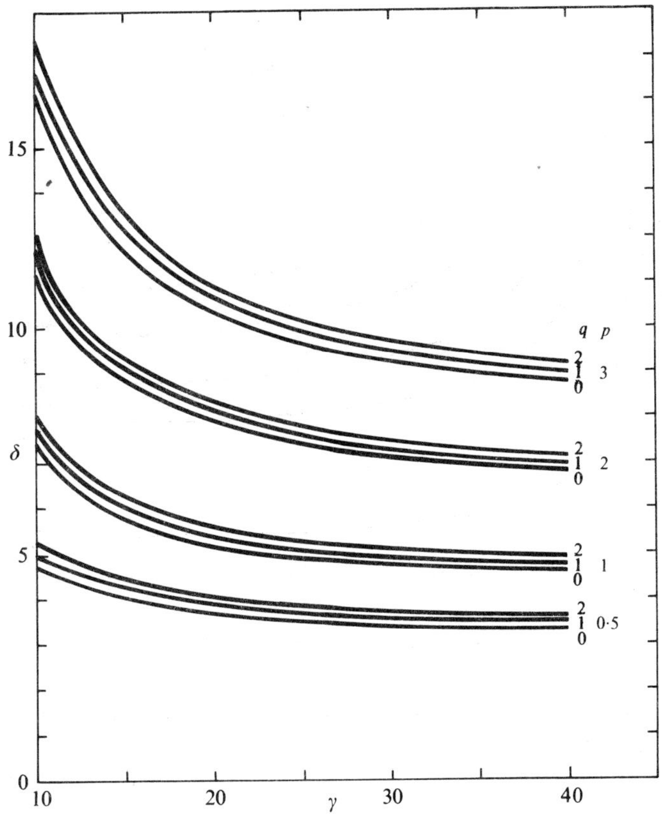

FIG. 6.6. Locus of conditions that just ensure uniqueness for four orders of reaction ($p = 0{\cdot}5$, 1, 2, 3) and the three symmetric shapes.

as a function of γ; the product $\beta\gamma$ must be less than this critical value if the solution is to be unique for all values of the Thiele modulus.

Villadsen and Michelsen (1972), using the very careful method of computation described in Sections 4.1.2 and 4.1.3, have found the boundaries of the regions in the β, γ which define precisely the maximum number of solutions to the first-order reaction in the sphere. These are shown in Fig. 6.7 amplified by some calculations which Mercer had done independently at about the same time (Mercer, 1973). For (β, γ) in the region $\Gamma_{2m-1} \cup \Gamma_{2m}$ there are at most $(2m-1)$ solutions; Γ_{2m-1} is the open region between the curves Γ_{2m} and Γ_{2m+2}. These curves Γ_{2m} are not truly hyperbolic, though they are approximated by the form $\beta\gamma = K_m(1+\beta)$. For example, Villadsen shows that K_1 varies from 4·38 at $\beta = 0{\cdot}07$ to 4·19 as $\beta \to \infty$. However the asymptote as $\beta \to \infty$ may be calculated by considering the form of the equation for large

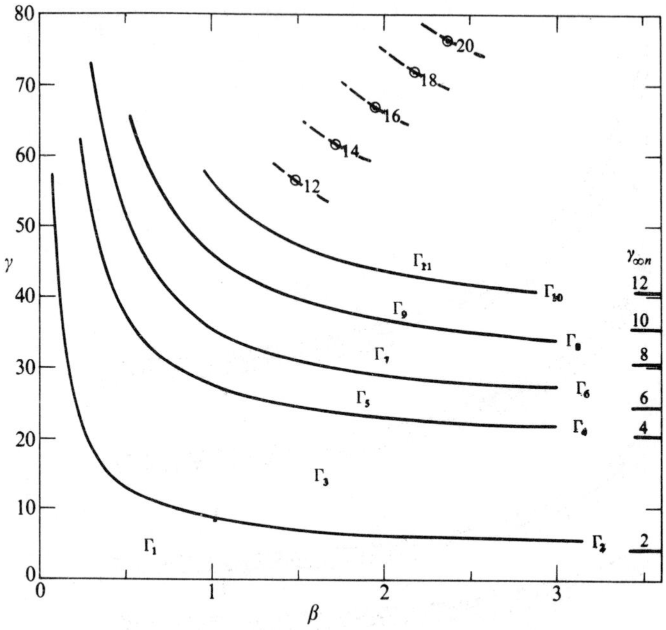

FIG. 6.7. Regions of multiplicity n in the (β, γ)-plane.

β. The asymptote $\gamma_{\infty n}$ to which the curves Γ_n go has been calculated by Villadsen and Michelsen:

n	2	4	6	8	10	12
$\gamma_{\infty n}$	4·188	20·153	24·170	30·301	35·308	40·622

The complex nature of the problem may be recognized from the intricacy of Fig. 6.8 which shows the relation between the Thiele modulus and the maximum value of $w(0)$ for five points (β, γ) in Γ_1, Γ_3, Γ_5, Γ_7, and Γ_9 respectively. It is seen that there is a range of ϕ, often quite small, in which the maximum multiplicity n can be attained. It is flanked by intervals in which $(n-2)$, $(n-4)$, ... solutions are to be found. The transformation of Villadsen and Michelsen (Section 4.1.3) whereby every point on the line $(\beta+1)/\gamma$ = constant can be obtained from one calculation is of the greatest value.

The model in which the catalyst pellet is at a uniform temperature, which is determined by the overall heat transfer to the pellet has been considered in various places, notably Sections 3.10 and 4.1.1, and it was remarked that there can be more than one solution to the equation for the temperature of the particle. This equation is

$$\eta\phi^2 = (\mu/\beta)(v_s - 1), \tag{6.26}$$

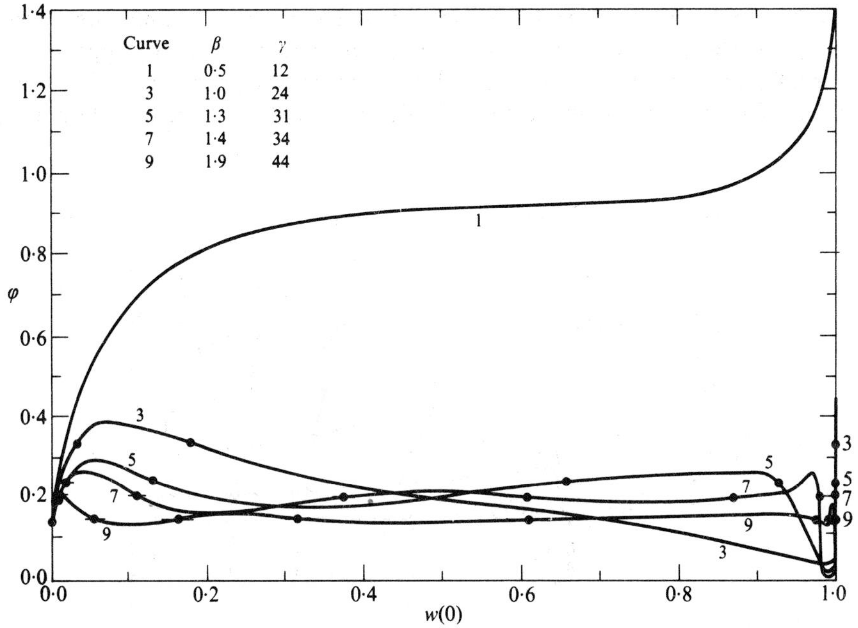

FIG. 6.8. The value of $w(0)$ as a function of Thiele modulus for various multiplicities. The marked points on any given curve show the various possible values of $w(0)$ for the same ϕ. N.B. All curves have $w = 1$ for vertical asymptote.

where η is the effectiveness factor calculated for the Robin problem. Thus for the sphere we would have the equation

$$\frac{3\nu(\phi\coth\phi-1)}{\nu+\phi\coth\phi-1}=\frac{\mu}{\beta}(v_s-1), \tag{6.27}$$

where

$$\phi^2=\phi_f^2\exp[\gamma(1-1/v_s)].$$

McGreavy and Thornton (1970*a*), in their study of this case, chose parameters which would allow an easy presentation of the results on uniqueness. Letting

$$v_s/\gamma=w, \quad \text{and} \quad \bar{\phi}^2=\phi_f^2\exp\gamma \tag{6.28}$$

be the unknown pellet temperature and Thiele modulus respectively, gives the equation

$$\frac{RT_f}{E}=\frac{1}{\gamma}=w-\left(\frac{\beta}{\mu\gamma}\right)\frac{3\nu(\phi\coth\phi-1)}{\nu-1+\phi\coth\phi}, \qquad \phi^2=\bar{\phi}^2\exp(-1/w). \tag{6.29}$$

In this equation $(\beta/\mu\gamma)$ can be regarded as independent of T_f and, like v and $\bar{\phi}$, can be chosen as a constant for the whole reactor. Thus T_f only appears on the left hand side of the equation and the solution w depends on T_f, $\bar{\phi}$, v, and $(\beta/\mu\gamma)$. Figure 6.9 shows the regions of uniqueness and multiplicity in the plane

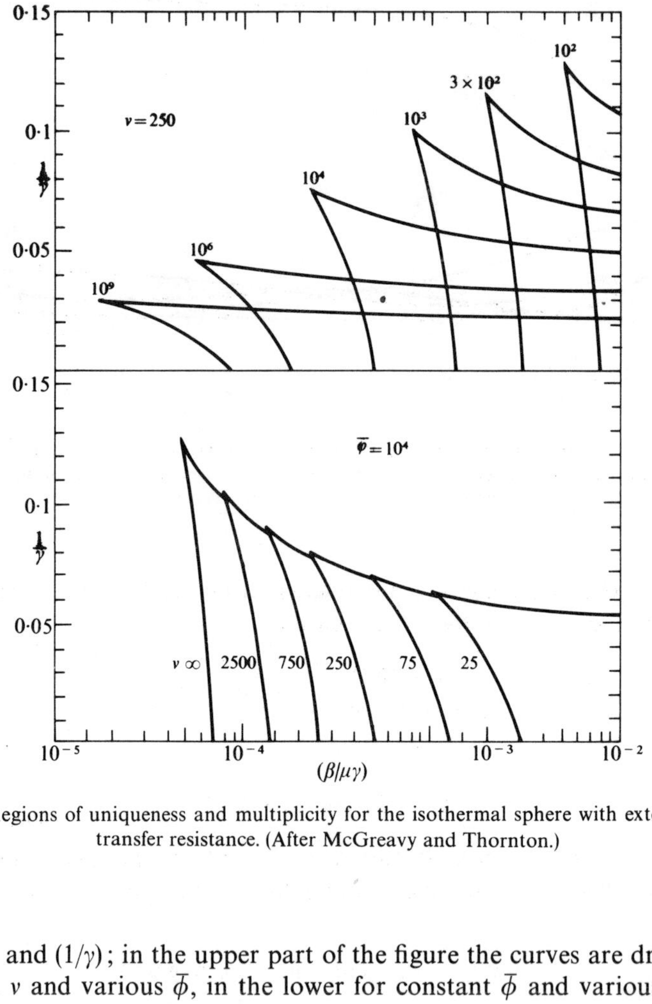

FIG. 6.9. Regions of uniqueness and multiplicity for the isothermal sphere with external heat transfer resistance. (After McGreavy and Thornton.)

of $(\beta/\mu\gamma)$ and $(1/\gamma)$; in the upper part of the figure the curves are drawn for constant v and various $\bar{\phi}$, in the lower for constant $\bar{\phi}$ and various v. The solution will be unique if the parameters lie between the arms of the curves. Such a diagram can be used to see when any pellet in a reactor is capable of having more than one steady state. The curves in these figures are very reminiscent of those for the stirred tank which were published by Aris and Regenass (1965).

6.2. Existence theorems

We shall attempt to summarize in this section some of the results that have been proved concerning the existence and character of the solutions of eqns (6.1) with boundary conditions given by (6.2) or (6.3). For the Dirichlet problem, or when $\mu = \nu$ in the Robin problem, they may be reduced to a single equation by the fact that

$$\beta u + v = 1 + \beta \tag{6.30}$$

in the steady state. It is convenient to set

$$w = 1 - u = (v - 1)/\beta, \tag{6.31}$$

so that

$$\nabla^2 w + \phi^2 R(1 - w, 1 + \beta w) = 0 \quad \text{in } \Omega \tag{6.32}$$

$$\frac{\partial w}{\partial \nu} + \mu w = 0 \quad \text{on } \partial\Omega \tag{6.33}$$

and $0 \leqslant w \leqslant 1$. When $R(u, v)$ is an increasing function of u and v,

$$P(w) = R(1 - w, 1 + \beta w), \tag{6.34}$$

will be a decreasing function of w if $\beta \leqslant 0$. If $\beta > 0$, we have seen in Fig. 2.3, where $P(w) = F(1 + \beta w)$, that $P(w)$ can take a variety of forms depending on the other kinetic parameters involved. We shall use our standard notation when applying various theorems to the diffusion reaction problem, but may quote results or describe theorems in the notation of their authors.

A theorem of Serrin, Douglas and Dupont (1971) ensures the uniqueness of the solution in certain cases corresponding broadly to endothermic reactions and isothermal reactions of order not less than one. Their theorem (2, corollary) concerns the elliptic differential equation in divergence form with

$$Lu = \sum \frac{\partial}{\partial x_i} A_i\left(x_i, u, \frac{\partial u}{\partial u_i}\right) = f(x, u)$$

in Ω and $u(x) = g(x)$ on $\partial\Omega$. This is applicable to our problem, and indeed includes cases of variable diffusion coefficients, if $f(x, u) = -\phi^2 P(u)$. Then, if $f(x, u)$ is non-decreasing in its second argument and is Lipschitz continuous in u for all x in any compact subset of Ω, there is at most one solution of the Dirichlet problem which is twice continuously differentiable in $\bar{\Omega}$. Because $f = -\phi^2 P$, the conditions of the theorem will hold for any P which is non-increasing and Lipschitz continuous. This will be the case for endothermic and isothermal reactions for which $P(1 - u) = O(u^p)$, $(p \geqslant 1)$, as $u \to 0$. We have already seen that with reactions of order less than one the continuity of the second derivative can break down.

Two theorems of Keller's (1969*b*) can be combined to give not only a proof of existence but also a constructive method for obtaining maximal and minimal solutions, that is solutions $\hat{w}(\boldsymbol{\rho})$ and $\check{w}(\boldsymbol{\rho})$ such that any other solution $w(\boldsymbol{\rho})$ satisfies

$$\check{w}(\boldsymbol{\rho}) \leqslant w(\boldsymbol{\rho}) \leqslant \hat{w}(\boldsymbol{\rho}). \tag{6.35}$$

The application of Keller's theorems requires us to define certain constants which we can most easily illustrate by taking the most awkward form $P(w)$ we have encountered. This is shown in the upper part of Fig. 6.10 and $f(w) = -\phi^2 P(w)$ is shown in the lower part. In applying Keller's work we note that we are specializing his equation $Lu = f(x, u)$, $B(u) = g(x, u)$ to the case where L is the Laplacian (rather than uniformly elliptic operator), f is independent of position, $g \equiv 0$ and the coefficients in $B(u) \equiv b_0(x)u + b_1(x)(\partial u/\partial v)$ are constant. It follows that we could entertain a more general problem with the diffusion coefficient varying throughout Ω and the mass transfer coefficient varying on $\partial\Omega$. Our problem is thus

$$Lw \equiv \nabla^2 w = f(w) \quad \text{in } \Omega, \qquad Bw \equiv vw + \frac{\partial w}{\partial v} = 0 \quad \text{on } \partial\Omega, \tag{6.36}$$

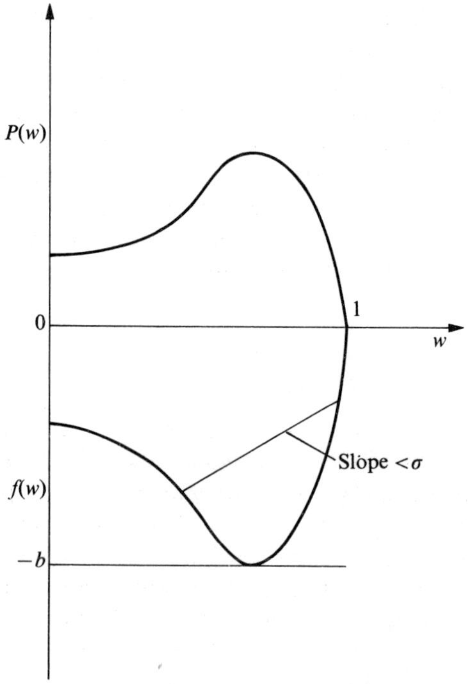

FIG. 6.10. The function $f(w)$ in relation to $P(w)$.

which is a straightforward specialization of Keller's general situation. We note that the slope of any chord of the curve $f(w)$ is bounded above by σ for all points in the range $0 \leqslant w \leqslant 1$ ($\sigma = \Omega(x)$ in Keller's notation, $m = 0$, $M = 1$), and that $f(0) < 0$, $f(1) = 0$. Also the value of $f(w)$ is bounded in $[0, 1]$ by $-b = -\phi^2 \max P(w)$ and zero. We have drawn $f(w)$ with the exothermic reaction in mind, that is

$$f(w) = -\phi^2(1-w)^p \exp\left(\frac{\beta\gamma w}{1+\beta w}\right),$$

and note that the slope of any chord is bounded above by $f'(1)$. Thus p must not be less than 1 if $\sigma = f'(1)$ is to be finite. However a more general kinetic law could be used and we assume in general that $f(w) \in C_\alpha[0, 1]$, i.e. that it is Hölder-continuous for w in $[0, 1]$. Let $\hat{w}_0$ satisfy

$$\nabla^2 \hat{w}_0 = -b \quad \text{in } \Omega, \qquad B\hat{w}_0 = 0 \quad \text{on } \partial\Omega \tag{6.37}$$

and define the sequence $\hat{w}_n(\mathbf{\rho})$ by the equations

$$\nabla^2 \hat{w}_n - \sigma \hat{w}_n = f(\hat{w}_{n-1}) - \sigma \hat{w}_{n-1} \tag{6.38}$$

in Ω with $B\hat{w}_n = 0$ on $\partial\Omega$. Then Keller's theorem proves that there is at least one solution $w(\mathbf{\rho})$ of eqn (6.36) which belongs to $C_{\alpha+2}[\bar{\Omega}]$. Of all such solutions there is a maximal solution $\hat{w}(\mathbf{\rho})$ satisfying eqn (6.35) and given by the limit of the sequence $\hat{w}_n(\mathbf{\rho})$ of solutions of eqns (6.37) and (6.38),

$$\hat{w}(\mathbf{\rho}) = \lim_{n\to\infty} \hat{w}_n(\mathbf{\rho}). \tag{6.39}$$

Moreover this is a monotonic decreasing sequence. Similarly if a sequence of functions $\check{w}_n(\mathbf{\rho})$ is defined by

$$\begin{aligned} L\check{w} &= 0, \qquad B\check{w}_0 = 0, \\ L\check{w}_n - \sigma\check{w}_n &= f(\check{w}_{n-1}) - \sigma\check{w}_{n-1}, \qquad B\check{w}_n = 0 \end{aligned} \tag{6.40}$$

then $\check{w}_n$ is a monotonic increasing sequence converging to $w(\mathbf{\rho})$ a minimal solution satisfying eqn (6.35). If the solution is unique the maximal and minimal solutions are identical.

Keller also shows that if f is monotone increasing and $\hat{w}_0$ is used to start the iteration scheme

$$Lw_n = f(w_{n-1}), \qquad Bw_n = 0, \tag{6.41}$$

and w_1 thus found is not negative, then the successive iterates will form an alternating sequence converging to the unique solution,

$$0 \leqslant w_1 \leqslant w_3 \leqslant \dots \leqslant w(\mathbf{\rho}) \leqslant \dots \leqslant w_2 \leqslant \hat{w}_0.$$

Similarly if w_1 is generated by eqn (6.41) from $\check{w}_0$ and $w_1 \leqslant 1$ then the successive iterates form an alternating sequence.

Cohen and Laetsch (1970) have used the same approach as Keller and shown that the solution exists and is unique if $P(w)$ is Hölder-continuous and continuously differentiable for $w \geqslant 0$ and $(P(w)/w)' < 0$ in $0 \leqslant w \leqslant 1$. Because the solution is bounded between 0 and 1, the function $P(w)$ can be continued beyond these bounds in any way that is necessary: Cohen and Laetsch do this by merely continuing the tangents at the end points.

Returning to questions of existence Kuiper (1971) has proved some very general results that would apply to the Dirichlet problem even if $P(w)$ were discontinuous. For our purposes however it assures us that for a continuous function $P(w)$ there exists an unbounded continuum of solutions from $\phi = 0$ to ∞. By this we mean that for each value of ϕ there will be a solution of $\nabla^2 w + \phi^2 P(w) = 0$ in Ω with $w = 0$ on $\partial\Omega$ which will belong to a suitable Banach space. In particular $\phi = 0$, $w = 0$ is a solution and this is connected to solutions for arbitrarily large values of ϕ. The theorem does not have anything to say about multiplicity. Hence if we were to picture the spectrum of solutions, Γ, in the plane of ϕ and $\|w\|$ (where $\|\,.\,\|$ is the norm appropriate to the Banach space) as in Fig. 6.11, we are assured of the continuity of Γ but

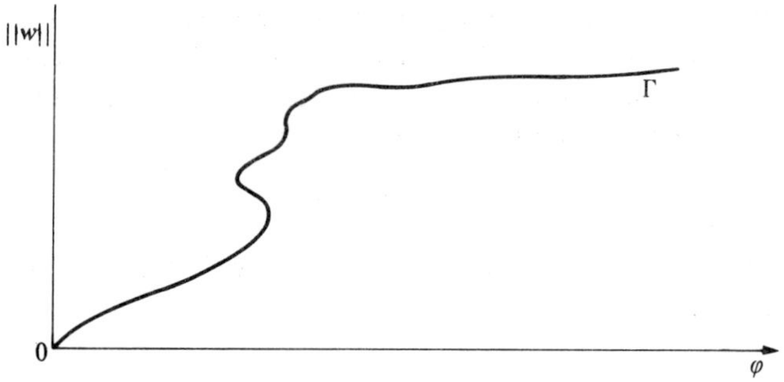

FIG. 6.11. The spectrum of solutions.

are not given any information about the number of intersections there may be of Γ with a vertical line through ϕ. In fact $1-u(0)$ in Fig. 6.8 will serve as $\|w\|$, and the complexity of the curve Γ becomes clear.

Amann (1972*b*) has shown that there are conditions when at least three solutions must exist. From Keller's theorem it is known that maximal and minimal solutions exist. Suppose that they are distinct and that $P(w)$ considered as a mapping in the space of continuous functions on $\bar{\Omega}$ has Fréchet derivatives belonging to $C^{\alpha}(\bar{\Omega})$ at $w = \check{w}$ and $w = \hat{w}$, then the boundary value problem

$$\nabla^2 w + \phi^2 P(w) = 0 \quad \text{in } \Omega,$$

$$\frac{\partial w}{\partial \nu} + \nu w = 0 \quad \text{on } \partial\Omega$$

has at least one more distinct solution $\bar{w}(\rho)$, $\check{w}(\rho) < \bar{w}(\rho) < \hat{w}(\rho)$, provided the boundary value problem

$$\begin{aligned} \nabla^2 v + \phi^2 P'(w) v &= 0 \quad \text{in } \Omega, \\ \frac{\partial v}{\partial \nu} + \nu v &= 0 \quad \text{on } \partial\Omega \end{aligned} \tag{6.42}$$

has only the trivial solution for $w = \check{w}$ and $w = \hat{w}$.

Gavalas (1968) was the first to apply index methods and other modern techniques to the existence problem. He showed that, when Ω is a sphere, the Dirichlet and Robin problems had solutions. Indeed, except when the derived equations analogous to (6.42) had nontrivial solutions, there must be an odd number of solutions. This is entirely analogous to Amann's result and could no doubt be generalized to any reasonable shape of Ω.

6.3. Sufficient conditions for uniqueness for all ϕ

We should ask first of all whether there are conditions which ensure the uniqueness of the solution whatever the value of the Thiele modulus. These will be conditions on the nature of the rate expression and can to some extent be read out of the theorems quoted in the last section. Thus endothermic and isothermal reactions of first or higher order will have a unique steady state. Also the theorem of Cohen and Laetsch shows that the condition $(P(w)/w)' < 0$ ensures uniqueness. Let us now prove this result using the method of Luss (1971), though staying with the form of the equation in w, namely

$$\nabla^2 w + \phi^2 P(w) = 0 \quad \text{in } \Omega, \tag{6.43}$$

with

$$\nu w + \frac{\partial w}{\partial \nu} = 0 \quad \text{on } \partial\Omega. \tag{6.44}$$

Suppose that there are two solutions of this equation w_1 and w_2, and w_1 is the minimal solution; i.e. $w_2 \geqslant w_1$. Then by Green's formula,

$$\iiint_{\Omega} (w_1 \nabla^2 w_2 - w_2 \nabla^2 w_1) \, \mathrm{d}\Upsilon = \iint_{\partial\Omega} \left(w_1 \frac{\partial w_2}{\partial \nu} - w_2 \frac{\partial w_1}{\partial \nu} \right) \mathrm{d}\Sigma,$$

and the second integral is identically zero by eqn (6.44). But we can substitute for the Laplacians from eqn (6.43) and then

$$\iiint_{\Omega} w_1 w_2 \{Q(w_1) - Q(w_2)\} \, \mathrm{d}\Upsilon = 0,$$

where

$$Q(w) = P(w)/w. \tag{6.45}$$

If $Q(w)$ is a non-increasing function, however, $Q(w_1) \geqslant Q(w_2)$, and, since w_1 and w_2 are never negative, this relation can only be satisfied if $w_1 = w_2$.

We recognize from the form of $P(w)$ given in Fig. 6.10 that the condition that $Q(w)$ should be non-increasing is the same as the condition that the tangent at the point of inflection should not pass below the origin. For the first-order reaction eqn (2.165) gives the condition for this as

$$\beta\gamma = \delta < 4(1+\beta), \tag{6.46}$$

and eqn (6.24) shows how close this is to the numerical values for the slab. For the pth-order reaction no neat closed form can be found. For an isothermal reaction with no autocatalytic overtones $R(u)$ is an increasing function so that $P(w)$ will be a decreasing function of w. It follows that the solution will be unique in such a case. $P'(w)$ will also be negative for an endothermic reaction and again uniqueness is ensured.

The only commonly occurring form of isothermal kinetic expression which can give multiple solutions is the substrate inhibited kinetics or Langmuir–Hinshelwood law with a quadratic denominator,

$$R(u) = \frac{u(1+\kappa)^2}{(\kappa+u)^2}.$$

In this case

$$\frac{P(w)}{w} = \frac{1-w}{w}\frac{(1+\kappa)^2}{(1+\kappa-w)^2},$$

and

$$Q'(w) = -\frac{(2w^2-3w+1+\kappa)(1+\kappa)^2}{w^2(1+\kappa-w)^3}.$$

It follows that if $8\kappa \geqslant 1$ uniqueness is assured. A similar analysis of the form

$$R(u) = \frac{1+\kappa+(1/\iota)}{(1+\kappa/u+u/\iota)}$$

shows that a sufficient condition for uniqueness is

$$\kappa > (1-\iota)^3/27\iota.$$

Jackson (1972*b*) has given sufficient conditions for the uniqueness of the Robin problem in the case of symmetric solutions in symmetric bodies. As was shown in Chapter 2, the equations for this case could be reduced to a single equation in either u or v at the expense of incorporating the surface value in the reaction rate expression. Thus from eqns (2.152–4) we have

$$u = 1 - \frac{\mu}{\nu\beta}(v_s - 1) - \frac{1}{\beta}(v - v_s), \tag{6.47}$$

with

$$\frac{1}{\rho^q}\frac{d}{d\rho}\left(\rho^q \frac{dv}{d\rho}\right) + \beta\phi^2 F(v; v_s) = 0, \tag{6.48}$$

subject to

$$\frac{dv}{d\rho} = 0 \quad (\rho = 0), \tag{6.49}$$

$$\frac{dv}{d\rho} + \mu v = \mu \quad (\rho = 1), \tag{6.50}$$

where

$$F(v; v_s) = R(u, v)$$

with u expressed in terms of v by eqn (6.47). Jackson divides the problem into two stages, treating first the Dirichlet problem with $v_s = v(1)$ given, and then finding the v_s which satisfies eqn (6.50). We will follow his procedure after first transforming to a form with homogeneous boundary conditions.

Let

$$\alpha = \nu/\mu, \tag{6.51}$$

$$v_s = 1 + \alpha\beta\tilde{w}, \qquad v = v_s + \beta w, \tag{6.52}$$

and

$$P(w; \tilde{w}) = R(1 - w - \tilde{w}, 1 + \beta w + \alpha\beta\tilde{w}). \tag{6.53}$$

For any fixed $\tilde{w}$ the range of w is from 0 to $(1-\tilde{w})$, while the range of $\tilde{w}$ is $(0, 1)$. The point $(w, \tilde{w})$ therefore lies in the triangular region D, shown in the lower part of Fig. 6.12. This is the image in the $(w, \tilde{w})$-plane of the admissible regions of the (v, v_s)-plane shown above and embraces both the case $\alpha > 1$ and $\alpha < 1$. The case of $\alpha = 1$ has been covered already. We note that, since $u = 1 - w - \tilde{w}$, the diagonal side of the triangle corresponds to $u = 0$. With these substitutions we have a Dirichlet problem for $w(\rho)$,

$$\begin{gathered}(\rho^q w')' + \rho^q \phi^2 P(w; \tilde{w}) = 0,\\ w'(0) = 0, \qquad w(1) = 0.\end{gathered} \tag{6.54}$$

This has to be solved for an arbitrary $\tilde{w}$ so that to be precise we should write $w = w(\rho; \tilde{w})$. Then $\tilde{w}$ is found from the boundary condition (6.50)

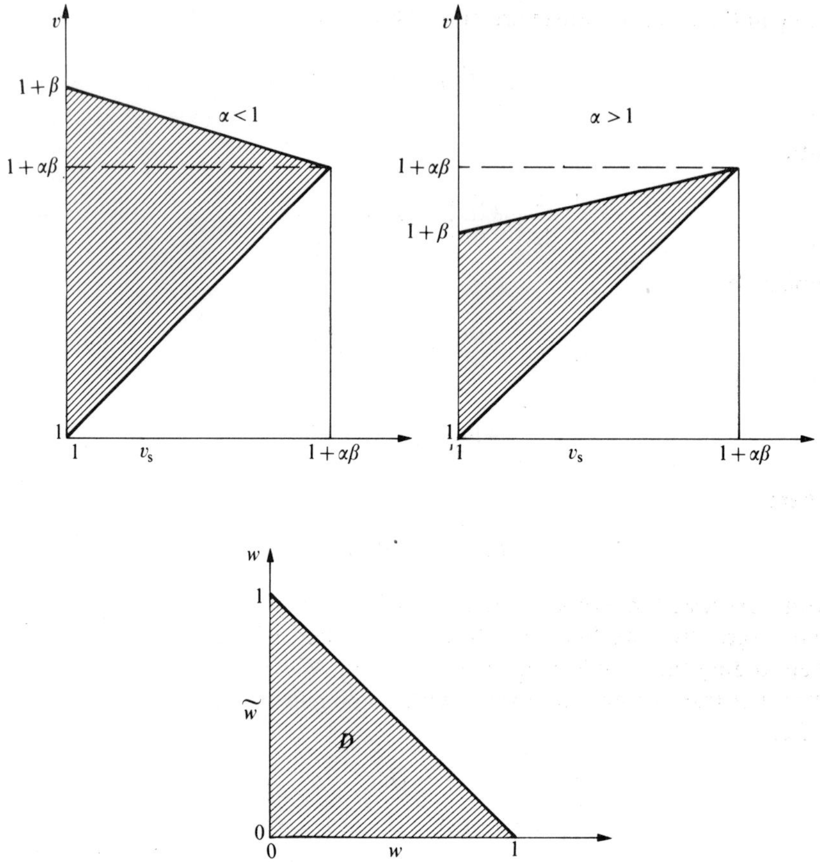

FIG. 6.12. The domain of variables for the symmetric Robin problem.

which in terms of w and $\tilde{w}$ is

$$g(\tilde{w}) \equiv w'(1\,;\tilde{w}) + v\tilde{w} = 0. \tag{6.55}$$

The uniqueness of the solution to the Robin problem eqns (6.48–50) will be assured if the solution to the Dirichlet problem (6.54) is unique and there is only one solution to eqn (6.55) in $0 \leqslant \tilde{w} \leqslant 1$. The latter condition may be difficult to establish and Jackson points out that we may have to be content with a further sufficient condition, namely that $g(\tilde{w})$ should be monotonic.

Now precisely the same argument as has been used before shows that the Dirichlet problem will have a unique solution, if

$$Q(w\,;\tilde{w}) = P(w\,;\tilde{w})/w$$

is monotonic decreasing in $0 \leqslant w \leqslant 1-\tilde{w}$; that is

$$Q'(w\,;\tilde{w}) < 0. \tag{6.56}$$

To discuss the monotonicity of $g(\tilde{w})$ we define (as in Sections 3.4.6 and 4.2.5) a function $\lambda(\rho)$ by

$$\begin{gathered}(\rho^q\lambda')' + \rho^q\phi^2 P_w(w\,;\tilde{w})\lambda = 0,\\ \lambda'(0) = 0, \qquad \lambda(1) = 1,\end{gathered} \tag{6.57}$$

where $P_w = \partial P/\partial w$ is evaluated at the solution $w(\rho)$ of eqns (6.54) and (6.55). For convenience, let a dot denote the partial derivative with respect to $\tilde{w}$, e.g. $\dot{w}' = \partial w'/\partial\tilde{w} = \partial(\mathrm{d}w/\mathrm{d}\rho)/\partial\tilde{w}$. Then $\dot{w}(\rho)$ satisfies

$$\begin{gathered}(\rho^q\dot{w}')' + \rho^q\phi^2 P_w\dot{w} = -\rho^q\phi^2\dot{P},\\ \dot{w}'(0) = 0, \qquad \dot{w}(1) = 0\end{gathered} \tag{6.58}$$

and the usual procedure of multiplying eqn (6.57) by $\dot{w}$ and eqn (6.58) by λ, subtracting and integrating leads to

$$\dot{w}'(1) = -\phi^2\int_0^1 \rho^q\dot{P}\lambda\,\mathrm{d}\rho. \tag{6.59}$$

Hence

$$\dot{g} = \frac{\mathrm{d}g}{\mathrm{d}\tilde{w}} = \nu - \phi^2\int_0^1 \rho^q\dot{P}\lambda\,\mathrm{d}\rho. \tag{6.60}$$

If P_w is bounded $\lambda(\rho)$ cannot change sign and so is always positive. It follows that the only sufficient condition giving a monotonic $g(\tilde{w})$ is

$$\dot{P} = \frac{\mathrm{d}}{\mathrm{d}\tilde{w}}P(w\,;\tilde{w}) < 0 \tag{6.61}$$

throughout D. Equations (6.56) and (6.61) are sufficient conditions for uniqueness for all values of ϕ.

If we apply this to the pth-order irreversible reaction with

$$R(u, v) = u^p\exp\left(\gamma - \frac{\gamma}{v}\right),$$

p137

i.e.

$$P(w\,;\tilde{w}) = (1-w-\tilde{w})^p\exp\left(\gamma - \frac{\gamma}{1+\beta w+\alpha\beta\tilde{w}}\right), \tag{6.62}$$

we have

$$P_w = -R_u + \beta R_v, \qquad \dot{P} = -R_u + \alpha\beta R_v.$$

Since $R_u = pR_w$, $R_v = \gamma R/v^2$,

$$Q' = \frac{w(-R_u + \beta R_v) - R}{w^2} = -\left\{w\left(p - \beta\gamma\frac{u}{v^2}\right) + u\right\}\frac{R}{uw^2} \tag{6.63}$$

and

$$\dot{P} = -\left\{p - \alpha\beta\gamma\frac{u}{v^2}\right\}\frac{R}{u}. \tag{6.64}$$

Also $w \geqslant 0$, $u \geqslant 0$ and $(u/v^2) \leqslant 1$ so that sufficient conditions for both Q' and $\dot{P}$ to be negative are

$$\alpha\beta\gamma < p, \qquad \beta\gamma < p. \tag{6.65}$$

These conditions are a little weak in the sense that when $\alpha = 1$, $p = 1$ they reduce to $\beta\gamma < 1$ thus requiring P_w, or $F'(v)$, to be monotonic. This is a more stringent requirement than Luss's $Q' < 0$ in the case $\alpha = 1$. There is no way of avoiding $\alpha\beta\gamma < p$ and if $\alpha \geqslant 1$ this is more stringent than $\beta\gamma < p$. However if $\alpha < 1$ there is some incentive for seeing if the crude estimates used in finding conditions for $Q' < 0$ cannot be refined. We may note in passing that the conditions (6.65) are always satisfied by an endothermic reaction, for β is then negative.

To examine the behaviour of $Q(w; \tilde{w})$ more closely, we appeal to the transformation given in Section 2.5.4, namely, if

$$P(w; \tilde{w}; p, \alpha, \beta, \gamma) = (1 - w - \tilde{w})^p \exp\left(\gamma - \frac{\gamma}{1 + \beta w + \alpha\beta\tilde{w}}\right)$$

and (6.66)

$$P(w; p, \beta, \gamma) = (1 - w)^p \exp\left(\gamma - \frac{\gamma}{1 + \beta w}\right),$$

we have

$$\begin{aligned} P(w; \tilde{w}; p, \alpha, \beta, \gamma) &= P(\tilde{w}; p, \alpha\beta, \gamma)P(w'; p, \beta', \gamma') \\ &= P(\tilde{w}'; p, \alpha\tilde{\beta}', \tilde{\gamma}')P(w; p, \beta, \gamma) \end{aligned} \tag{6.67}$$

where

$$\beta' = \beta\frac{1 - \tilde{w}}{1 + \alpha\beta\tilde{w}}, \qquad \gamma' = \gamma\frac{1}{1 + \alpha\beta\tilde{w}}, \qquad w' = \frac{w}{1 - \tilde{w}},$$

and (6.68)

$$\tilde{\beta}' = \beta\frac{1 - w}{1 + \beta w}, \qquad \tilde{\gamma}' = \gamma\frac{1}{1 + \beta w}, \qquad \tilde{w}' = \frac{\tilde{w}}{1 - w}.$$

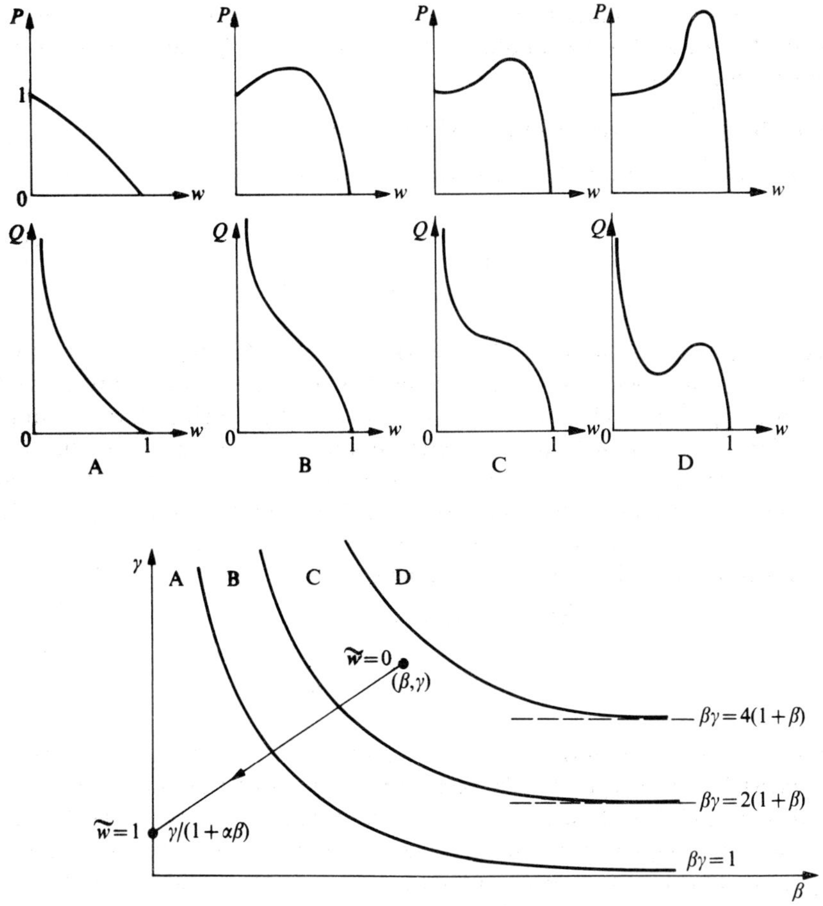

FIG. 6.13. The path of (β', γ') and transformations of P.

Now we know the forms that $P(w'; p, \beta', \gamma')$ can take when (β', γ') lies in the several parts of the (β, γ)-plane, and eqns (6.68) tell us that as $\tilde{w}$ increases from 0 to 1 the point (β', γ') will move down the straight line from (β, γ) to $(0, \gamma/(1+\alpha\beta))$. The situation for a first-order reaction is shown in Fig. 6.13 where sketches of the qualitative behaviour of P and Q identified with the different regions are given. This qualitative behaviour is unaffected by the first factor in eqn (6.67), and it follows that if (β, γ) lies below the curve $\beta\gamma = 4(1+\beta)$ then so must (β', γ') for all $\tilde{w}$. Thus $Q'(w; \tilde{w}) < 0$ for all $\tilde{w}$ if $Q'(w; 0) < 0$ and, for $\underline{p} = 1$, we may write

$$\beta\gamma < \min(1/\alpha, 4(1+\beta))$$

as the sufficient conditions for uniqueness. When $p \neq 1$ there is no longer a simple formula for the condition of monotonicity of Q, but a similar curve can be calculated.

6.4. Bounds on the region of multiplicity

By (ϕ_b, ϕ^b) we mean the open interval of ϕ within which the equation

$$\begin{aligned} \nabla^2 w + \phi^2 P(w) &= 0 \quad \text{in } \Omega \\ \frac{\partial w}{\partial \nu} + \nu w &= 0 \quad \text{on } \partial\Omega \end{aligned} \tag{6.69}$$

has more than one solution. At $\phi = \phi_b$ and $\phi = \phi^b$ there are usually two solutions, one of which is the confluence of two solutions which are distinct for $\phi > \phi_b$ or $\phi < \phi^b$. The exception to this can be imagined for the Robin problem. If in Fig. 4.8 the parameters μ and ν were so adjusted that the (η, ϕ)-curve had a vertical bitangent at $\phi = \phi_b$, there would be a transition from one to five solutions, and two of the three solutions at $\phi = \phi_b$ would be the confluences of two pairs of solutions distinct for $\phi > \phi_b$. It is because of this confluence that the derived equation

$$\begin{aligned} \nabla^2 v + \phi^2 P'(w) v &= 0 \quad \text{in } \Omega \\ \frac{\partial v}{\partial \nu} + \nu v &= 0 \quad \text{on } \partial\Omega \end{aligned} \tag{6.70}$$

has a nontrivial solution when $w = w_b$ or w^b, the confluent solutions of eqns (6.69) at $\phi = \phi_b$ or ϕ^b. This accounts for the conditions in the theorems of Amann and Gavalas which assert that there are an odd number of solutions if the derived equation has only trivial solutions. To see how this is given by the confluence let $w_1(\boldsymbol{\rho})$ and $w_2(\boldsymbol{\rho})$ be the two distinct solutions for $\phi > \phi_b$ which tend to $w_b(\boldsymbol{\rho})$ as $\phi \to \phi_b + 0$. Then, by subtraction of eqns (6.67) for $w = w_1$ and $w = w_2$,

$$\begin{aligned} \nabla^2(w_1 - w_2) + \phi^2\{P(w_1) - P(w_2)\} &= 0 \quad \text{in } \Omega \\ \frac{\partial}{\partial \nu}(w_1 - w_2) + \nu(w_1 - w_2) &= 0 \quad \text{on } \partial\Omega \end{aligned} \tag{6.71}$$

Let

$$\|w_1 - w_2\| = \varepsilon$$

and

$$W_\varepsilon = (w_1 - w_2)/\varepsilon,$$

and write

$$P(w_1) - P(w_2) = \left[\int_0^1 P'\{w_1 + \tau(w_2 - w_1)\}\, d\tau\right](w_1 - w_2) \tag{6.72}$$

Then W_ε does not vanish since $\|W_\varepsilon\| = 1$ for all ε and

$$\nabla^2 W_\varepsilon + \phi^2\left[\int_0^1 P'\left\{w_2 + \tau(w_1 - w_2)\right\} \mathrm{d}\tau\right] W_\varepsilon = 0. \tag{6.73}$$

Now let $\phi \to \phi_\mathrm{b} + 0$ so that $w_1 \to w_2 \to w_\mathrm{b}$ and $\varepsilon \to 0$. Then if W_b is the limit of W_ε as $\varepsilon \to 0$,

$$\begin{gathered} \nabla^2 W_\mathrm{b} + \phi_\mathrm{b}^2 P'(w_\mathrm{b}) W_\mathrm{b} = 0, \\ \frac{\partial W_\mathrm{b}}{\partial v} + \nu W_\mathrm{b} = 0. \end{gathered} \tag{6.74}$$

The continuity of this operation ensures that $W_\mathrm{b}(\boldsymbol{\rho})$ does not vanish identically since it is the limit of functions of unit norm, and hence the derived equation has a nontrivial solution.

In studying the bounds on ϕ_b and ϕ^b we are not interested in situations for which the solution is unique. Hence the function $P(w)$ is as shown on the left of Fig. 6.14 and $Q(w) = P(w)/w$, the slope of the line from the origin, is

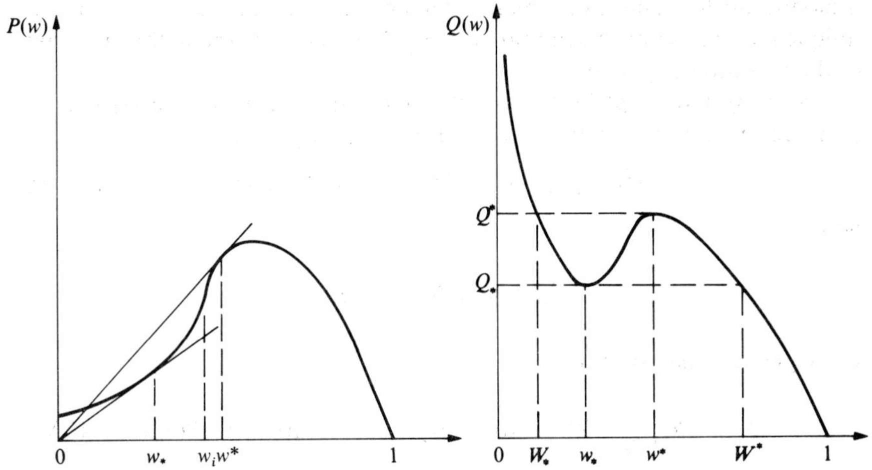

FIG. 6.14. The relation between the $P(w)$ and $Q(w)$ curves showing the important abscissae.

not monotonic. This is shown at the right of the figure. Let w_* and w^* denote the abscissae of the two points of tangency of the line from the origin to the $P(w)$ curve. Then $Q(w)$ has a minimum at w_* and a maximum at w^*; also

$$Q_* = Q(w_*) = P'(w_*) \quad \text{and} \quad Q^* = Q(w^*) = P'(w^*). \tag{6.75}$$

Also let W_* and W^* denote the points (other than w^* and w_*) at which

$$Q(W_*) = Q^* \quad \text{and} \quad Q(W^*) = Q_* \tag{6.76}$$

and denote by w_i the inflection point of the $P(w)$ curve and by P_i' its slope there. Now the argument on uniqueness given in the previous paragraph turned on the fact that if w_1 is the minimal solution and w_2 any other solution

$$\iiint_{\Omega} \{Q(w_1)-Q(w_2)\}w_1 w_2 \,\mathrm{d}\Upsilon = 0, \tag{6.77}$$

and the monotonicity of $Q(w)$ prevented the integrand from changing sign. If there are two solutions there must be some 'intertwining' in the sense that $Q(w_1)-Q(w_2)$ changes sign (for w_2-w_1 is always positive) and it is this which the monotonicity suffices to prevent. But by the same token, if we could ensure that all of the solutions satisfy

$$w(\boldsymbol{\rho}) \leqslant w_*, \tag{6.78}$$

they would be confined to the monotonic part of the $Q(w)$ curve and hence there could be only one. But the solution for $\phi = 0$ is $w \equiv 0$, and if

$$\omega(\phi) = \max_{\boldsymbol{\rho}\in\Omega}\{w(\boldsymbol{\rho})\} \tag{6.79}$$

is monotonic for sufficiently small ϕ, then we can be sure that the solution is unique for small ϕ provided that $\omega < w_*$. Let us first examine this question of the monotonicity of ω.

Let $w_1(\boldsymbol{\rho})$ and $w_2(\boldsymbol{\rho})$ be two solutions of eqns (6.43) and (6.44) corresponding to ϕ_1 and ϕ_2 with $\phi_1 < \phi_2$. If $v = w_1 - w_2$, then

$$\nabla^2 v + \{\phi_1^2 P(w_1) - \phi_2^2 P(w_2)\} = 0 \quad \text{in } \Omega \tag{6.80}$$

with

$$\frac{\partial v}{\partial \nu} + \nu v = 0 \quad \text{on } \partial\Omega.$$

But eqn (6.77) may be written

$$\nabla^2 v + \phi_1^2\left[\int_0^1 P'\{w_2 + \tau(w_1 - w_2)\}\,\mathrm{d}\tau\right] v = (\phi_2^2 - \phi_1^2)P(w_2). \tag{6.81}$$

Also a theorem on the maximum principle (Protter and Weinberger (1967), Chapter 2, Theorem 10) states that if $\nabla^2 v + f(\boldsymbol{\rho})v > 0$, but there exists a positive function $u(\boldsymbol{\rho})$ such that $\nabla^2 u + f(\boldsymbol{\rho})u < 0$, then v is negative in Ω. Identifying f with the coefficient of v on the left hand side of eqn (6.81) we see that $\nabla^2 v + fv > 0$ for $\phi_2 > \phi_1$ and we have only to find u. Consider u_1, the eigenfunction of

$$\nabla^2 u + \lambda^2 u = 0 \quad \text{in } \Omega \tag{6.82}$$

$$\frac{\partial u}{\partial \nu} + \nu u = 0 \quad \text{on } \partial\Omega \tag{6.83}$$

corresponding to the least eigenvalue λ_1^2. If w_1 and w_2 are less than w_i, the abscissa of the inflection point of $P(w)$, and if

$$\phi_1^2 \dot{P}_i' < \lambda_1^2, \tag{6.84}$$

then

$$\nabla^2 u_1 + f(\boldsymbol{\rho}) u_1 = -\{\lambda_1^2 - f(\boldsymbol{\rho})\} u < 0,$$

and by the maximum principle u_1 is positive in Ω. This shows that $u_1(\boldsymbol{\rho})$ will serve as the function required in the statement of the theorem, and therefore that $v(\boldsymbol{\rho}) = w_1(\boldsymbol{\rho}) - w_2(\boldsymbol{\rho})$ is negative. It follows, *a fortiori*, that

$$\omega(\phi_2) > \omega(\phi_1) \quad \text{for } \phi_2 > \phi_1 \tag{6.85}$$

provided that

$$\omega(\phi_2) < w_i. \tag{6.86}$$

Since the function $Q(w)$ is monotonic decreasing for $\omega < w^*$ and $\omega(\phi)$ is monotonic for $\omega < w_i$, we have the following condition for uniqueness. Let ϕ_* be the value of ϕ such that

$$\omega(\phi_*) = w_*, \tag{6.87}$$

then the solution is unique for $\phi \leqslant \phi_*$ and hence $\phi_b \geqslant \phi_*$. However this condition is not particularly useful since we would have to solve the equation to determine ϕ_*. Luss (1971) has shown how it may be translated into a more effective criterion. Consider the Poisson equation

$$\nabla^2 \tilde{u} + 1 = 0 \quad \text{in } \Omega, \tag{6.88}$$

$$\frac{\partial \tilde{u}}{\partial v} + v\tilde{u} = 0 \quad \text{on } \partial\Omega, \tag{6.89}$$

and let

$$U = \max_{\boldsymbol{\rho} \in \Omega} [\tilde{u}(\boldsymbol{\rho})]. \tag{6.90}$$

Since $\omega(\phi)$ is monotonic for $\omega < w_i$ we can determine a unique ψ_* such that

$$\omega(\psi_*) = W_* \tag{6.91}$$

and take

$$u_*(\boldsymbol{\rho}) = \psi_*^2 P(W_*) \tilde{u}(\boldsymbol{\rho}) \tag{6.92}$$

where $\tilde{u}(\boldsymbol{\rho})$ satisfies eqns (6.88) and (6.89). Then if $\phi < \psi_*$,

$$\nabla^2(u_* - w) = -\{\psi_*^2 P(W_*) - \phi^2 P(w)\} \leqslant 0,$$

and by the maximum principle $u_* > w$ in Ω. Thus the maximum value of

u_* is greater than that of v, or

$$U\psi_*^2 P(W_*) > W_*,$$

i.e.

$$\psi_*^2 > \frac{1}{U}\frac{W_*}{P(W_*)} = \frac{1}{UQ(W_*)} = \frac{1}{UP'(w^*)}. \tag{6.93}$$

It follows that

$$\phi_b^2 > \psi_*^2 > \{UP'(w^*)\}^{-1}. \tag{6.94}$$

An upper bound for ψ_* can also be found by comparing eqns (6.70) with eqns (6.82) and (6.83). For, by the same argument as has been used before,

$$\iiint_\Omega \{\phi^2 Q(w) - \lambda_1^2\} u_1 w \,\mathrm{d}\Upsilon = 0, \tag{6.95}$$

and there must be a part of Ω where the argument of the integral is negative. But setting $\phi = \psi_*$, which implies $Q(w) > Q^*$, gives

$$\psi_* < \lambda_1\{Q(w^*)\}^{-\frac{1}{2}} = \lambda_1\{P'(w^*)\}^{-\frac{1}{2}}. \tag{6.96}$$

The relative disposition of these bounds is shown in Fig. 6.15. ϕ_* is likely to be the closest lower bound for ϕ_b, but unfortunately its definition in

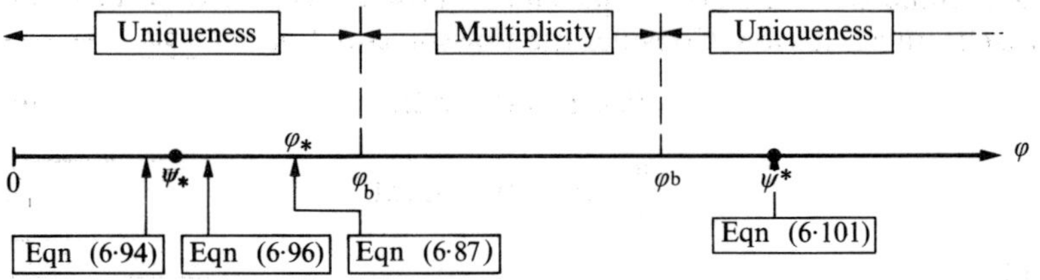

FIG. 6.15. The disposition of the various estimates.

eqn (6.87) requires the solution of the equation. ψ_* is a lower bound for ϕ_*, and hence for ϕ_b, and we can provide upper and lower bounds for ϕ_* by equations (6.94) and (6.96); these are obtained straight from the properties of $P(w)$ and do not require any knowledge of the solution of the equation. These two bounds on ψ_* are often quite close. Thus for the Dirichlet problem in the three symmetrical shapes of slab, cylinder, and sphere using the radius of half-thickness as characteristic length a, we have

q	0	1	2
$U^{-\frac{1}{2}}$	$2^{\frac{1}{2}} = 1{\cdot}414$	$4^{\frac{1}{2}} = 2$	$6^{\frac{1}{2}} = 2{\cdot}449$
λ_1	$\pi/2 = 1{\cdot}571$	$j_{01} = 2{\cdot}405$	$\pi = 3{\cdot}142$

For the Robin problem in the symmetric shapes

$$U^{-\frac{1}{2}} = \left(\frac{2(q+1)v}{v+2}\right)^{\frac{1}{2}} \tag{6.97}$$

and λ_1 is the smallest root of one of the equations

$$\begin{aligned} \lambda \tan \lambda &= v & q &= 0, \\ \lambda J_1(\lambda) &= vJ_0(\lambda) & q &= 1, \\ \lambda \cot \lambda &= 1-v & q &= 2, \end{aligned} \tag{6.98}$$

that is, of

$$\lambda J_{\frac{1}{2}(q+1)}(\lambda) = vJ_{\frac{1}{2}(q-1)}(\lambda).$$

These may be used to give a lower bound for ϕ_b in the case of the Robin problem with $\mu = v$.

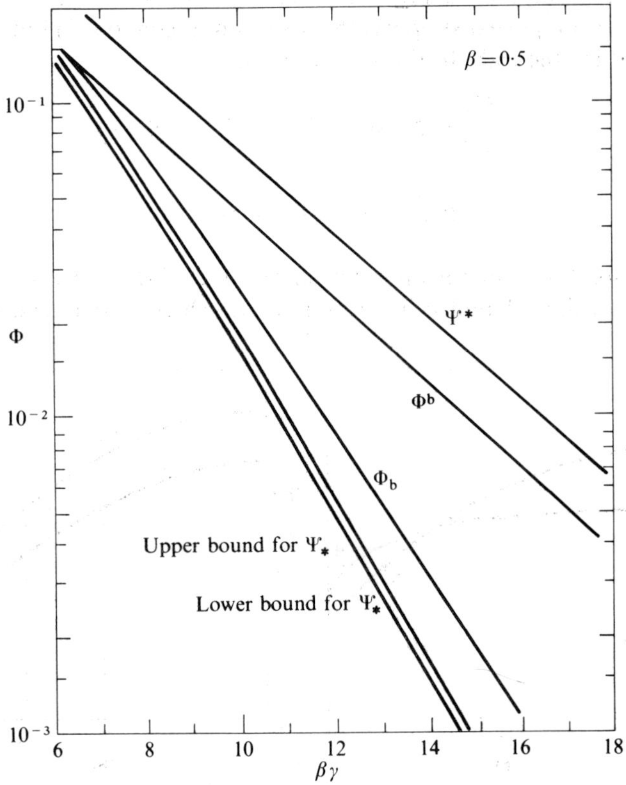

FIG. 6.16. The relation between the normalized bifurcation values and the bounds of ψ.

In Fig. 6.16 some comparisons are made between the exact value of Φ_b, i.e. the normalized value of ϕ_b, for the slab. Luss's bound, in the same normalization denoted by Ψ_*, is between the two parallel curves below Φ_b. Also shown on this figure are the upper bifurcation value Φ^* and Luss's upper bound for this, Ψ^*, to the consideration of which we now turn. Luss (1971) showed that if an upper bound for $\omega(\phi^b)$, the maximum value of the solution w^b, could be found then eqn (6.95) would yield an upper bound ψ^* for ϕ^b. For putting $\phi = \phi^b$ in eqn (6.95) shows that there must be a part of Ω in which the integrand is negative. Hence

$$(\lambda_1/\phi^b)^2 > \min_{\rho\in\Omega}[Q\{w(\rho)\}] = \min_{0\leqslant w\leqslant\omega(\phi^b)}[Q(w)]. \tag{6.99}$$

If it could be shown that

$$\omega(\phi^b) \leqslant W^* \tag{6.100}$$

we would then have

$$\phi^b < \psi^* = \lambda_i(\mathrm{Min}\ Q)^{-\frac{1}{2}} = \lambda_1(Q_*)^{-\frac{1}{2}} = \lambda_1\{P'(w_*)\}^{-\frac{1}{2}}. \tag{6.101}$$

It is ψ^* which is shown in normalized form Ψ^* in Fig. 6.16. Every calculation that has been reported supports the assertion in eqn (6.100) but it can only be proved for the slab. In the slab we have

$$\frac{d^2w}{d\rho^2}+\phi^2P(w) = 0, \quad (-1 \leqslant \rho \leqslant 1),$$

$$\pm\frac{dw}{d\rho}+\nu w = 0, \quad (\rho = \pm 1)$$

and two solutions cannot intersect, since if they did there would be a pair of points (such as A and B in Fig. 6.17(a)) at which w and $dw/d\rho$ would be

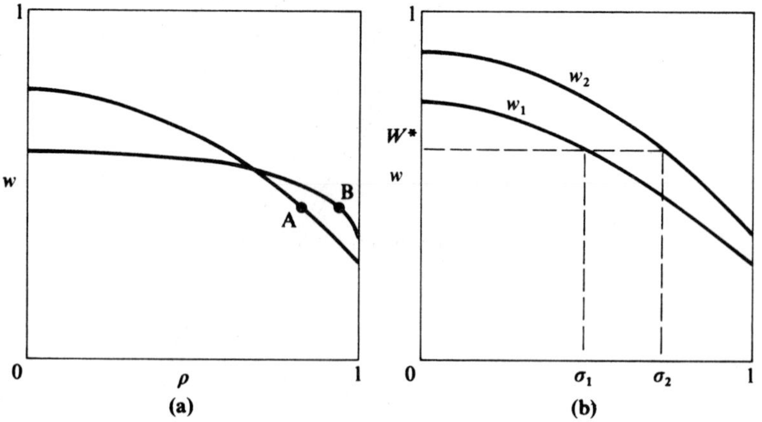

FIG. 6.17. (a) The two points at which w and $dw/d\rho$ are the same if w_1 and w_2 cross. (b) The definition of σ_1 and σ_2.

the same. But this would imply a crossing at a non-critical point of the trajectories in the phase plane of w and $dw/d\rho$, which is impossible. If we assume that there are two solutions for $\phi = \psi^*$, given by eqn (6.101), we shall arrive at a contradiction. For eqn (6.95) shows that if there were two solutions for this value of ϕ, say w_1 and w_2, then their maximum values, $w_1(0)$ and $w_2(0)$, would both have to be greater than W^*. Without loss of generality we can take $w_2(\rho) > w_1(\rho)$, since the solutions do not intersect, and choose σ_1 and σ_2 so that

$$w_1(\sigma_1) = w_2(\sigma_2) = W^*. \tag{6.102}$$

From Fig. 6.17(b) we see that $\sigma_2 > \sigma_1$. Denote the integrand in eqn (6.95) by

$$M_j(\rho) = [\phi^2 Q\{w_j(\rho)\} - \lambda_1^2]u_1 w_j \qquad j = 1, 2, \tag{6.103}$$

so that if $\phi = \psi^*$ then $M_j(\rho)$ is negative in $(0, \sigma_j)$, and positive in $(\sigma_j, 1)$. Thus we may write eqn (6.95) as

$$I_{ji} = \int_0^{\sigma_j} -M_j(\rho)\,d\rho = \int_{\sigma_j}^1 M_j(\rho)\,d\rho = I_{jo} \quad j = 1, 2, \tag{6.104}$$

the second suffix on the I denoting the inner or outer part of the integral. The integrands of all four integrals are positive. Since $Q(w)$ is monotonic decreasing if $w > W^*$, $-M_{2i} > M_{1i}$ and, because $\sigma_2 > \sigma_1$,

$$I_{2i} > I_{1i} \tag{6.105}$$

Also the outer integrals can be expressed

$$I_{jo} = \int_{\sigma_j}^1 M_j(\rho)\,d\rho = \int_{w_{js}}^{W^*} M_j(w)\left(-\frac{d\rho}{dw}\right) dw$$

and $w_{2s} > w_{1s}$, $(-dw_1/d\rho) < (-dw_2/d\rho)$. Hence

$$I_{1o} > I_{2o}. \tag{6.106}$$

But eqns (6.105) and (6.106) imply that

$$I_{1o} > I_{2o} = I_{2i} > I_{1i},$$

which contradicts eqn (6.104) for $j = 1$. Hence the assertion that there are two solutions for $\phi = \psi^*$ is false. Varma has pointed out that this still does not prove that $\phi^b < \psi^*$ and Luss has completed the proof as follows (both private communications). Assume that $\phi^b > \psi^*$ and define w^b by

$$\phi^b = \lambda_1\{Q(w^b)\}^{-\frac{1}{2}}$$

Then since $\phi^b > \psi^* = \lambda_1\{Q(W^*)\}^{-\frac{1}{2}}$, we must have $w^b > W^*$. But by using eqn (6.99) again we now have that any solutions must satisfy

$$w_i(0) > w^b > W^*.$$

Then, replacing W^* by w^{b} in eqn (6.102), the same argument shows that there cannot be two solutions. It follows that $\phi^{\mathrm{b}} < \psi^*$.

No proof has yet been found for the general region Ω. It is tempting to try and follow the same steps, taking $w_2(\boldsymbol{\rho})$ to be the maximal solution and assuming that there is another solution for $\phi = \psi^*$. Since the maximal solution does not intersect $w_1(\boldsymbol{\rho})$, the region Ω_2, within which $w_2(\boldsymbol{\rho}) \geqslant W^*$ certainly includes as a subregion the corresponding

$$\Omega_1 = \rho; \{\boldsymbol{\rho} \in \Omega, w_1(\boldsymbol{\rho}) \geqslant W^*\}.$$

The integral (7.95) can again be split into two parts and

$$I_{ji} = \iiint_{\Omega_j} -M_j \, \mathrm{d}\Upsilon = \iiint_{\Omega - \Omega_j} M_j \, \mathrm{d}\Upsilon = I_{jo}. \tag{6.107}$$

Again

$$\begin{aligned} I_{2i} = \iiint_{\Omega_2} -M_2 \, \mathrm{d}\Upsilon = \iiint_{\Omega_1} -M_2 \, \mathrm{d}\Upsilon + \iiint_{\Omega_2 - \Omega_1} -M_2 \, \mathrm{d}\Upsilon > \\ > \iiint_{\Omega_1} -M_1 \, \mathrm{d}\Upsilon = I_{1i} \end{aligned} \tag{6.108}$$

Let $\partial\Omega_{jw}$ denote the level surface $w_j(\boldsymbol{\rho}) = w$ and $g_j = \partial w_j/\partial n$ and $\mathrm{d}\Sigma_j$ be the normal gradient and surface element. Then

$$\begin{aligned} I_{1o} = \iiint_{\Omega - \Omega_1} M_1 \, \mathrm{d}\Upsilon = \int_0^{W^*} M(w) \, \mathrm{d}w \iint_{\partial\Omega_{1w}} \frac{\mathrm{d}\Sigma_1}{g_1} > \\ > \int_0^{W^*} M(w) \, \mathrm{d}w \iint_{\partial\Omega_{2w}} \frac{\mathrm{d}\Sigma_2}{g_2} = I_{2o} \end{aligned} \tag{6.109}$$

if

$$f_1(w) = \iint_{\partial\Omega_{1w}} \frac{\mathrm{d}\Sigma_1}{g_1} > \iint_{\partial\Omega_{2w}} \frac{\mathrm{d}\Sigma_2}{g_2} = f_2(w). \tag{6.110}$$

This last inequality seems highly plausible, but I do not see how to prove it. Certainly

$$\int_0^{W^*} f_1(w) \, \mathrm{d}w = \text{Volume of } (\Omega - \Omega_1) > \text{Volume of } (\Omega - \Omega_2) = \int_0^{W^*} f_2(w) \, \mathrm{d}w, \tag{6.111}$$

but unfortunately $M(w)$, though positive, is zero at w_* and W^*, and a change of sign of $(f_1 - f_2)$ near these points does not allow us to infer eqn (6.109) from (6.111).

Returning to the Robin problem for a symmetrical solution in a symmetrical body, we recall that Jackson (1972*b*) divides it into a Dirichlet problem, eqn (6.54), followed by an equation for $\tilde{w}$, eqn (6.55). The Dirichlet problem in w has a nonlinear term $P(w\,;\tilde{w})$ and any condition we find for its uniqueness must recognize that $\tilde{w}$ can have any value in [0, 1]. Thus the estimate of ϕ_{b} given by eqn (6.93) would ensure uniqueness for

$$\phi^2 < \min_{0<\tilde{w}<1} \psi_*^2(\tilde{w}), \tag{6.112}$$

where

$$\psi_*^2(\tilde{w}) = \{UP_w(w^*\,;\tilde{w})\}^{-1}, \tag{6.113}$$

and $w^* = w^*(\tilde{w})$ is the point at which $Q(w\,;\tilde{w})$ has a local maximum.

We are interested in the situation for which $Q(w\,;\tilde{w})$ is not monotonic for all $\tilde{w}$, so that the surfaces P and Q may look as shown in Figs. 6.18 and 6.19. If α is not too large, the maximum of $P_w(w^*\,;\tilde{w})$ with respect to $\tilde{w}$ will occur at $\tilde{w} = 0$, but there may be cases for which Q has an internal maximum. Using the relations given in eqn (6.67) this will occur when the equations

$$w'P_{w'}(w'\,;p,\beta',\gamma') = P(w'\,;p,\beta',\gamma')$$

and

$$P_{\tilde{w}'}(\tilde{w}'\,;p,\alpha\tilde{\beta}',\tilde{\gamma}') = 0$$

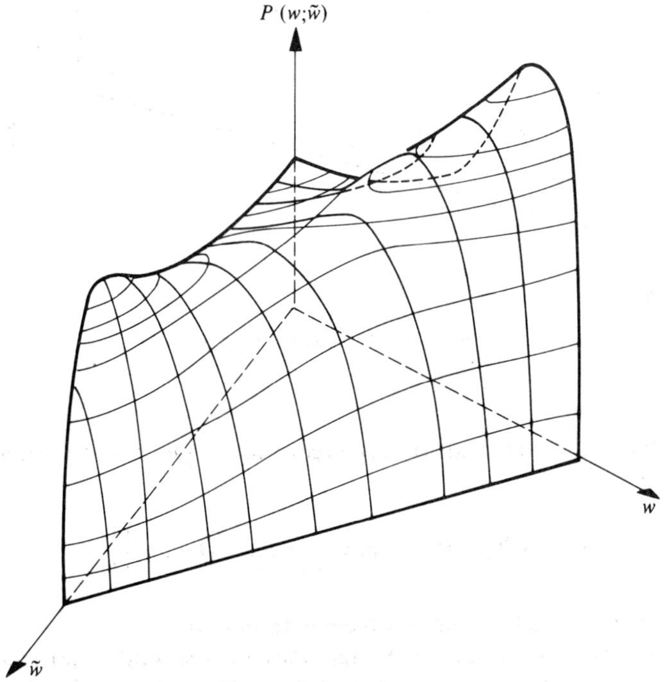

FIG. 6.18. The surface $P(w\,;\tilde{w})$.

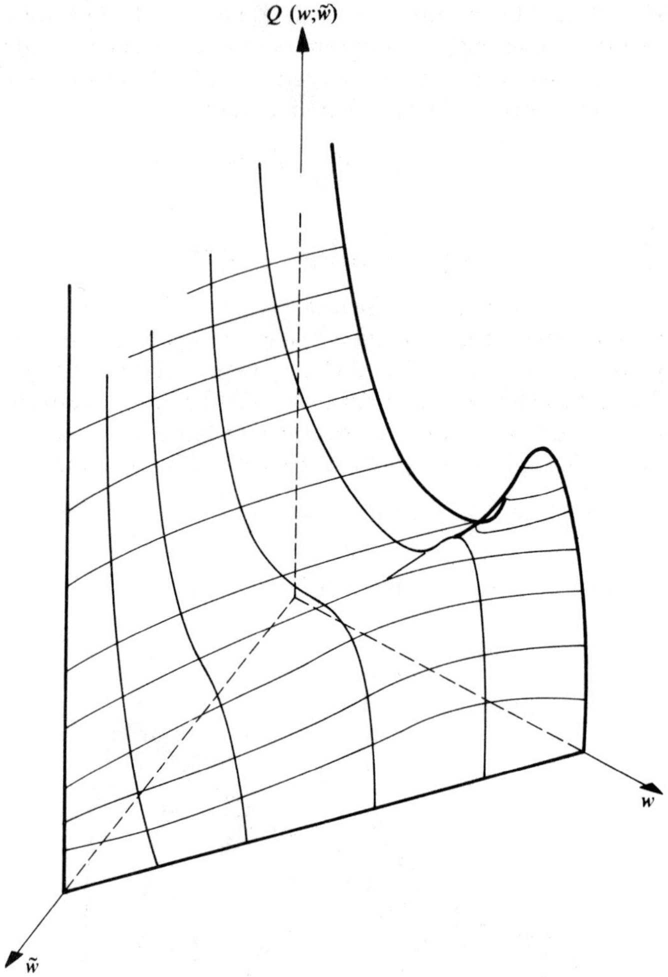

FIG. 6.19. The surface $Q(w;\tilde{w})$.

are simultaneously satisfied. In any event the uniqueness condition (6.112) can be written

$$\phi < \left\{2(q+1)\Big/\left(\max_{0\leqslant\tilde{w}\leqslant 1}\ \max_{0\leqslant w\leqslant 1-\tilde{w}} Q(w;\tilde{w})\right)\right\}^{\frac{1}{2}}, \tag{6.114}$$

for U is given by the Dirichlet problem with infinite v.

It is also clear from eqn (6.55) that Jackson's second requirement, the monotonicity of $g(\tilde{w})$, will be satisfied for sufficiently small values of ϕ.

If $\dot{P}(w;\tilde{w}) = \partial P/\partial \tilde{w}$ has a positive upper bound

$$E = \max_{0 \leq w \leq 1} \max_{0 \leq \tilde{w} \leq 1-w} \left\{ P_{\tilde{w}'}(w';p;\alpha\tilde{\beta}',\tilde{\gamma}')\frac{P(w;p,\beta,\gamma)}{1-w} \right\}. \tag{6.115}$$

Now if ϕ satisfies eqn (6.114) it must also satisfy the condition corresponding to eqn (6.96), i.e. with λ_1 rather than $\{2(q+1)\}^{\frac{1}{2}}$. Hence $\lambda(\rho)$ is positive and not greater than 1. It follows that $\dot{g}$ will be positive if

$$\phi < \{v(q+1)/E\}^{\frac{1}{2}}. \tag{6.116}$$

Hence uniqueness will be assured if ϕ is less than either of the bounds given in eqns (6.114) and (6.116). Jackson has pursued the details of the calculation further for the case of a first-order irreversible reaction and shows that the bounds so attained vary in their conservatism.

A lower bound may also be found for the trifurcation modulus, the value $\phi = \phi_t$ at which the (η, ϕ)-curve has an inflection point with vertical tangent when the other parameters of the problem are just right, e.g. for the exothermic reaction (β, γ) must lie on Γ_2. This inflection point represents the confluence of three solutions brought about by the passage of the other parameters through a critical constellation. To fix ideas let us consider the exothermic reaction where the situation for points such as A, B, and C in Γ_1, Γ_2, and Γ_3 respectively is as shown in the three diagrams of the lower part of Fig. 6.20. As (β, γ) moves from C to B, the two bifurcation values ϕ_b and ϕ^b coalesce at ϕ_t. Let $w_1(\mathbf{\rho})$, $w_2(\mathbf{\rho})$ and $w_3(\mathbf{\rho})$ be the three solutions at $\phi = \phi_t$ in situation C. Then

$$\begin{aligned} \nabla^2 w_j + \phi_t^2 P(w_j) &= 0 \quad \text{in } \Omega \\ \frac{\partial w_j}{\partial v} + v w_j &= 0 \quad \text{on } \partial\Omega \quad j = 1, 2, 3. \end{aligned} \tag{6.117}$$

Let $v_{ij} = (w_i - w_j)/\|w_i - w_j\|$ and $\bar{w}_{ij}$ be a value of w in the interval (w_i, w_j). Then subtracting the jth equation from the ith and using the mean-value theorem gives

$$\begin{aligned} \nabla^2 v_{ij} + \phi_t^2 P'(\bar{w}_{ij}) v_{ij} &= 0 \quad \text{in } \Omega, \\ \frac{\partial v_{ij}}{\partial v} + v v_{ij} &= 0 \quad \text{on } \partial\Omega. \end{aligned} \tag{6.118}$$

Further, let $u = (v_{32} - v_{21})/\|v_{32} - v_{21}\|$ and repeat the procedure with the equations for v_{32} and v_{21} giving

$$\begin{aligned} \nabla^2 u + \phi_t^2 P''(\bar{w}_{123}) u &= 0 \quad \text{in } \Omega, \\ \frac{\partial u}{\partial v} + v u &= 0 \quad \text{on } \partial\Omega, \end{aligned} \tag{6.119}$$

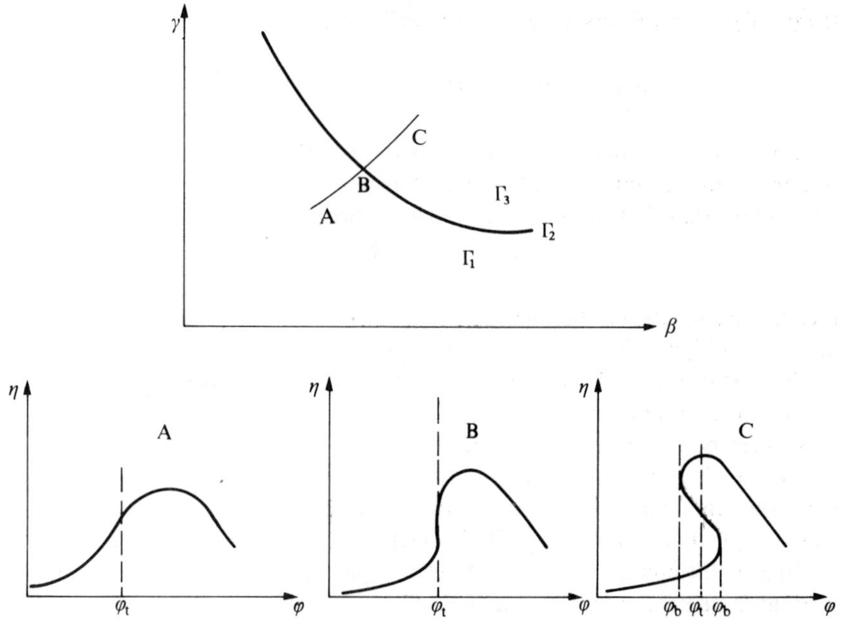

FIG. 6.20. The situation when passing from a region of uniqueness to a region of multiplicity in the space of the auxiliary parameters.

where $\bar{w}_{123}$ lies in the interval $(\bar{w}_{32}, \bar{w}_{21})$. If now the parameter point (β, γ) approaches B, w_1, w_2, and w_3 all approach w_t and hence pinch $\bar{w}_{ij}$ and $\bar{w}_{123}$ into w_t. The functions v_{32} and v_{21} approach a common function v which is a nontrivial solution of the derived equation

$$\nabla^2 v + \phi_t^2 P'(w_t) v = 0. \tag{6.120}$$

At the same time the second derived equation

$$\nabla^2 u + \phi_t^2 P''(w_t) u = 0 \tag{6.121}$$

also has a nontrivial solution.

It follows that both

$$\phi_t^2 \max_{0 \leqslant w \leqslant 1} [P'(w)] > \lambda_1^2$$

and

$$\phi_t^2 \max_{0 \leqslant w \leqslant 1} [P''(w)] > \lambda_1^2.$$

Thus

$$\phi_t^2 > \lambda_1^2 / \min\{P'(w_i), \max P''(w)\}. \tag{6.122}$$

6.4.1. *Approximate methods*

The analogy with the stirred tank which was developed earlier to render the multiplicity of steady states plausible, and the methods based on single point collocation which yield similar equations, can be used to get estimates of the bifurcation points. Their chief disadvantage is their unreliability in the sense that it is never clear whether an estimate of ϕ_b will lie above or below the true value, thus multiplicity may be predicted in some cases when the solution is unique and uniqueness in others for which there are really several solutions. Of course a one point collocation method may be improved by increasing the number of points, but this is just a way of computing the exact result to any desired degree of accuracy. However the approximate methods are so simple that they are worth exploring and when used with care can yield useful results. Villadsen and Stewart (1969) were the first to show the simplicity and usefulness of such methods and Padmanabhan and van den Bosch (1974*a*) have studied the question of their validity and accuracy. The 'linearization' described above in Section 6.1 was first given by Frank-Kamenetskii (1947) and has been extensively cultivated by Hlaváček and his coworkers (Hlaváček, 1970*a*; Hlaváček and Kubíček, 1970; Hlaváček, Kubíček and Caha, 1971). Luss and Lee (1971) have developed an alternative lumping procedure.

The argument for the stirred tank analogy used in the first section of this chapter for the Dirichlet problem can be extended to the Robin problem as follows. Let

$$\frac{\partial u}{\partial \tau} = \nabla^2 u - \phi^2 R(u, v), \qquad \mathscr{L}\frac{\partial v}{\partial \tau} = \nabla^2 v + \beta\phi^2 R(u, v) \quad \text{in } \Omega,$$

$$\frac{\partial u}{\partial \nu} + \nu u = \nu, \qquad \frac{\partial v}{\partial \nu} + \mu v = \mu \qquad \text{on } \partial\Omega,$$

and consider the linear problem

$$\begin{aligned} \nabla^2 W + \lambda^2 W &= 0 \quad \text{in } \Omega, \\ \frac{\partial W}{\partial \nu} + \kappa W &= 0 \quad \text{on } \partial\Omega. \end{aligned} \tag{6.123}$$

The linear problem has a denumerable sequence of eigenvalues of which $\lambda_1^2(\kappa)$ is the least. Then the approximation consists in replacing $\nabla^2 u$ by $-\lambda_1^2(\nu)(u-1)$ and $\nabla^2 v$ by $-\lambda_1^2(\mu)(v-1)$. This gives

$$\begin{aligned} \frac{\mathrm{d}u}{\mathrm{d}\tau'} &= 1 - u - \psi^2 R(u, v), \\ L\frac{\mathrm{d}v}{\mathrm{d}\tau'} &= 1 - v + \beta\psi^2 R(u, v), \end{aligned} \tag{6.124}$$

where

$$\tau' = \lambda_1^2(\nu)\tau, \qquad \psi = \phi/\lambda_1(\nu), \qquad L/\mathscr{L} = \beta'/\beta = \lambda_1^2(\nu)/\lambda_1^2(\mu). \tag{6.125}$$

For the symmetrical shapes of slab, cylinder and sphere, the $\lambda_1(\nu)$ are the least root of eqns (6.98). The steady state equations may be solved by setting

$$u = 1-w, \qquad v = 1+\beta' w, \qquad P(w) = R(1-w, 1+\beta' w), \tag{6.126}$$

so that

$$w = \psi^2 P(w) \tag{6.127}$$

In the one-point collocation method the partial differential equations are replaced by

$$\frac{du_1}{d\tau} = \frac{2\nu}{2+\nu\{1-\rho_{(1)}^2\}}(1-u_1) - \phi^2 R(u_1, v_1), \tag{6.128}$$

$$\mathscr{L}\frac{dv_1}{d\tau} = \frac{2\mu}{2+\mu\{1-\rho_{(1)}^2\}}(1-v_1) + \beta\phi^2 R(u_1, v_1), \tag{6.129}$$

where u_1 and v_1 are the values of u and v at the collocation point $\rho_{(1)}$. This point was originally given in eqn (4.39) as

$$\rho_{(1)}^2 = \frac{q+1}{q+5},$$

but Padmanabhan and van den Bosch suggest moving it from this value to $\rho_{(1)}^2 = 0{\cdot}5$ as ϕ increases. In any event we shall again get eqns (6.124) if we let $2\kappa/\{2+\kappa-\kappa\rho_{(1)}^2\}$ play the role of $\lambda_1^2(\kappa)$ for $\kappa = \nu$ or $\kappa = \mu$.

Now eqn (6.127) lends itself to a graphical exposition and, as Villadsen and Stewart (1969) have shown, can give a very reasonable estimate of the region of multiplicity in certain cases. For the first-order irreversible the curves given by Aris and Regenass (1965) may be adapted to the purpose. For if

$$R(u, v) = u\,e^{\gamma - (\gamma/v)}$$

the equations (6.124) in the steady state can be written (in their notation) as

$$m(y - y^*) = z = \alpha\,e^{-1/y}/(1+\alpha\,e^{-1/y})$$

where

$$m = (\gamma/\beta'), \qquad y^* = (1/\gamma), \qquad \alpha = \psi^2\,e^{\gamma}, \qquad y = v/\gamma.$$

Figure 6.21, which is just Regenass' Fig. 2, may be used to find the limits of ϕ between which the stirred tank analogue predicts multiplicity for any given β, γ, μ and ν. For, using either eqn (6.125) or its collocation equivalent, the values of these parameters give the coordinates (γ/β') and $(1/\gamma)$ of a point

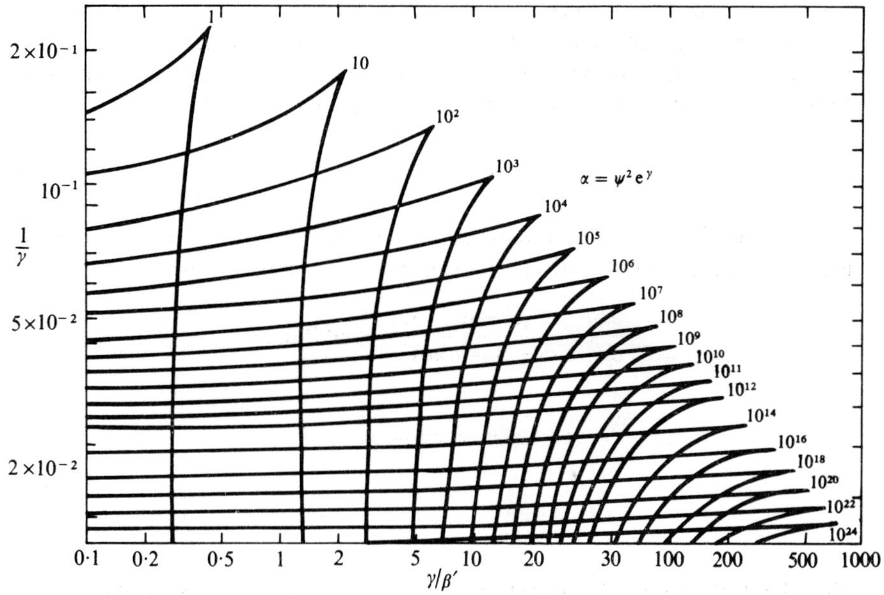

FIG. 6.21. Chart for determining the limits of uniqueness of the equivalent stirred tank.

in this figure. The values of $\psi^2 e^\gamma$ that correspond to the limits of the region of multiplicity can then be estimated from the two values of α that pass through the point. From these values the corresponding estimates of ϕ_b and ϕ^b can be obtained.

The equivalent stirred tank approximation gives a sufficient condition for uniqueness for all ϕ which is the same as in Section 6.3, namely that $Q'(w) = \{P(w)/w\}'$ should be negative. But here of course P has the parameters β' and γ, so that for a first-order irreversible reaction it is

$$\beta'\gamma < 4(1+\beta')$$

or

$$\beta\gamma < 4\left\{\frac{\lambda_1^2(\mu)}{\lambda_1^2(\nu)}+\beta\right\}, \tag{6.130}$$

where $\lambda_1^2(\kappa)$ can be interpreted either as an eigenvalue or by the collocation approximation. Using the latter with $\rho_{(1)}^2 = (q+1)/(q+5)$ gives

$$\beta\gamma < 4\left(\frac{q+5+2\nu}{q+5+2\mu}\,\frac{\mu}{\nu}+\beta\right), \tag{6.131}$$

a neat formula with suitable limiting forms but unknown reliability.

It is clear that the reduction of the problem to an equivalent stirred tank can only lead to the detection of two bifurcation values, since the stirred tank can have at the most three steady states. On the other hand, it is known that even in the slab the Robin problem has five steady states for suitable $\phi, \beta, \gamma, \mu, \nu$. Padmanabhan and van den Bosch (1973*a*) follow Cresswell and Patterson (1971) in approximating the profile by a parabola in

$$\sigma = \frac{\rho-\omega}{1-\omega}, \tag{6.132}$$

where ω is the radius within which the concentration of the reactant is identically zero. The equation for the concentration becomes

$$\frac{d^2u}{d\sigma^2}+\frac{q(1-\omega)}{\omega+(1-\omega)\sigma}\frac{du}{d\sigma} = \phi^2R(u, v), \tag{6.133}$$

with

$$\frac{du}{d\sigma}+\nu(1-\omega)u = \nu(1-\omega) \quad \text{at } \sigma = 1$$

and

$$u = \frac{du}{d\sigma} = 0 \quad \text{at } \sigma = 0. \tag{6.134}$$

Taking a one-point collocation approximation and writing an equation for u_1, the value of $u(\sigma)$ at $\sigma_{(1)}$ gives a more complicated but more accurate equation for large values of ϕ. Padmanabhan and van den Bosch show that, for the slab problem, good accuracy can be obtained using this approximation and $\sigma_{(1)}^2 = 0{\cdot}5$ for large ϕ, and the classical one-point collocation with $\rho_{(1)}^2 = 0{\cdot}2$ for small ϕ. As is always the case with such methods, the accuracy is unreliable, the estimate being sometimes above and sometimes below the true value. Comparisons for this and several other problems are made by Hlaváček and Kubíček (1970, 1971*b*) and Hlaváček, Kubíček, and Caha (1971).

6.5. Multiple reactions

Some of the earliest work on the uniqueness and stability of reaction in catalyst particles touched the general problem of multiple reactions, and we should at this point take note of the approach that Gavalas used in it. We shall base our exposition on Gavalas' splendid little monograph (1968), though his papers in 1966 and 1967 should not be overlooked. Taking the case of constant and independent diffusion coefficients we have the system of equations

$$\Delta_i\nabla^2u_i+\sum_{j=1}^{R}\alpha_{ji}\phi_j^2R_j(u, v) = 0 \tag{6.135}$$

$$\nabla^2v+\sum_{j=1}^{R}\beta_j\phi_jR_j(u, v) = 0 \tag{6.136}$$

in Ω. In the argument of R_j the variable u stands for the set of all the u_j, $i = 1, 2, \ldots S$. The boundary conditions are

$$\frac{1}{\nu_i}\frac{\partial u_i}{\partial \nu} + u_i = u_{if}, \qquad \frac{1}{\mu}\frac{\partial v}{\partial \nu} + v = 1. \tag{6.137}$$

Let $\mathscr{G}(\boldsymbol{\rho}, \boldsymbol{\sigma}; \kappa)$ be the Robin function for the Laplacian in the region Ω, with homogeneous boundary conditions

$$\frac{1}{\kappa}\frac{\partial \mathscr{G}}{\partial \nu} + \mathscr{G} = 0$$

on $\partial\Omega$. Then

$$u_i(\boldsymbol{\rho}) = u_{if} + \sum_{j=1}^{R} \frac{\alpha_{ji}\phi_j^2}{\Delta_i} \iiint_{\Omega} \mathscr{G}(\boldsymbol{\rho}, \boldsymbol{\sigma}; \nu_i) R_j(u(\boldsymbol{\sigma}), v(\boldsymbol{\sigma}))\, \mathrm{d}\Upsilon_\sigma \tag{6.138}$$

$$v(\boldsymbol{\rho}) = 1 + \sum_{j=1}^{R} \beta_j \phi_j^2 \iiint_{\Omega} \mathscr{G}(\boldsymbol{\rho}, \boldsymbol{\sigma}, \mu) R_j(u(\boldsymbol{\sigma}), v(\boldsymbol{\sigma}))\, \mathrm{d}\Upsilon_\sigma. \tag{6.139}$$

For example, the Robin function for a sphere is

$$\mathscr{G}_2(\rho, \sigma; \kappa) = \begin{cases} \dfrac{1}{4\pi}\left(\dfrac{1}{\sigma} + \dfrac{1}{\kappa} - 1\right), & 0 \leqslant \rho \leqslant \sigma, \\[2ex] \dfrac{1}{4\pi}\left(\dfrac{1}{\rho} + \dfrac{1}{\kappa} - 1\right), & \sigma \leqslant \rho \leqslant 1. \end{cases} \tag{6.140}$$

For the Dirichlet problem the Green's functions, as the Robin function is then called, are the same in the two equations (6.138) and (6.139). For independent reactions we can write

$$\begin{aligned} u_i(\boldsymbol{\rho}) &= u_{is} + \sum_{j=1}^{R} \frac{\alpha_{ji}}{\Delta_i} w_j(\boldsymbol{\rho}), \\ v(\boldsymbol{\rho}) &= 1 + \sum_{j=1}^{R} \beta_j w_j(\boldsymbol{\rho}), \end{aligned} \tag{6.141}$$

and

$$w_j(\boldsymbol{\rho}) = \phi_j^2 \iiint_{\Omega} \mathscr{G}(\boldsymbol{\rho}, \boldsymbol{\sigma}) P_j(w(\boldsymbol{\sigma}))\, \mathrm{d}\Upsilon_\sigma, \quad j = 1, \ldots, R, \tag{6.142}$$

where

$$P_j(w) = P_j(w_1, \ldots, w_R) = R_j\left(u_s + \sum \frac{\alpha_{ji}}{\Delta_i} w_j, 1 + \Sigma \beta_j w_j\right). \tag{6.143}$$

If $\mathbf{w}$ denotes the vector of the w and $\mathscr{H}$ the nonlinear transformation defined by eqns (6.142), then the solution is a fixed point of $\mathscr{H}$ in the space of real continuous functions defined on Ω with zero values on the boundary. It is a completely continuous operator (see Gavalas, (1968) p. 65) and we write

$$\mathbf{w} = \mathscr{H}\mathbf{w}. \tag{6.144}$$

In the Robin problem we gain little by trying to reduce the dimensionality and it is better to put

$$\left.\begin{aligned} w_i(\boldsymbol{\rho}) &= u_i(\boldsymbol{\rho}) - u_{if} \\ w_{S+1}(\boldsymbol{\rho}) &= v(\boldsymbol{\rho}) - 1 \end{aligned}\right\}, \tag{6.145}$$

and define $\mathscr{H}$ as the nonlinear transformation in the integrals of eqns (6.138) and (6.139). This leads to the same form as eqn (6.144), but with a larger vector $\mathbf{w}$ and a more complicated operator. In both cases, however, this formulation leads to an existence proof by showing that the vector fields $\mathbf{I}$ and $\mathbf{I}$–$\mathscr{H}$ are homotopic.

The Frechet derivative $\mathscr{H}'$ of the operator $\mathscr{H}$ at a solution $\mathbf{w}$ is such that

$$\mathscr{H}(\mathbf{w} + \varepsilon\mathbf{v}) = \mathscr{H}(\mathbf{w}) + \varepsilon\mathscr{H}'(\mathbf{w})\mathbf{v} + o(\varepsilon) \tag{6.146}$$

so that for eqn (6.144) the derived equation is the linear equation

$$\mathbf{v} = \mathscr{H}'\mathbf{v}. \tag{6.147}$$

For the Dirichlet problem this gives

$$v_j(\boldsymbol{\rho}) = \iiint\limits_{\Omega} \mathscr{G}(\boldsymbol{\rho}, \boldsymbol{\sigma}) \sum_{k=1}^{R} \phi_j^2 \left(\frac{\partial P_j}{\partial w_k}\right) v_k(\boldsymbol{\sigma})\, \mathrm{d}\Upsilon_\sigma, \tag{6.148}$$

or $\mathscr{H}'$ is the matrix operator whose (j, k)th element is

$$(\mathscr{H}')_{jk} = \iiint\limits_{\Omega} \mathscr{G}(\boldsymbol{\rho}, \boldsymbol{\sigma})\phi_j^2(\partial P_j/\partial w_k) \cdot \mathrm{d}\Upsilon_\sigma \tag{6.149}$$

A similar, but more complicated, matrix can be defined for the full Robin problem.

The proof of uniqueness can proceed on one of two lines both of which show that the solution is unique when the ϕ_j are all sufficiently small. In the first place all the derivatives of the reaction rate functions are bounded and, though they may be large, we have only to make the ϕ_j sufficiently small for the operator $\mathscr{H}$ to be a contraction operator. As an example we may revert to the single exothermic reaction and note that then

$$\mathscr{H}w(\boldsymbol{\rho}) = \phi^2 \iiint\limits_{\Omega} \mathscr{G}(\boldsymbol{\rho}, \boldsymbol{\sigma}) P(w(\boldsymbol{\sigma}))\, \mathrm{d}\Upsilon_\sigma \tag{6.150}$$

and hence

$$|\mathscr{H}w_1(\boldsymbol{\rho}) - \mathscr{H}w_2(\boldsymbol{\rho})| = \phi^2 \iiint_\Omega \mathscr{G}(\boldsymbol{\rho}, \boldsymbol{\sigma})|P(w_1) - P(w_2)|\,\mathrm{d}\Upsilon \leqslant$$

$$\leqslant \phi^2 \max_{0 \leqslant w \leqslant 1} |P'(w)| \iiint_\Omega \mathscr{G}(\boldsymbol{\rho}, \boldsymbol{\sigma})|w_1(\boldsymbol{\sigma}) - w_2(\boldsymbol{\sigma})|\,\mathrm{d}\Upsilon.$$

If $\|w\| = \max_{\boldsymbol{\rho}\in\Omega} |w(\boldsymbol{\rho})|$, we have

$$\|\mathscr{H}w_1 - \mathscr{H}w_2\| \leqslant \phi^2 U|P'_{\max}|\,\|w_1 - w_2\|,$$

where U is the maximum of the solution of $\nabla^2 u = 1$ in Ω. $\mathscr{H}$ is therefore a contraction operator if $\phi^2 < \{U|P'_{\max}|\}^{-1}$, and though this is not nearly as good a bound in general as that given by eqn (6.93), it shows that uniqueness can be proved in this way. The other line of attack is to say that if a bifurcation solution exists, then the derived equation must have a non-trivial solution. More comprehensively (see Gavalas (1968) p. 99, or Krasnosel'skii (1964) p. 136) if $\mathbf{w}$ is a fixed point of eqn (6.144) and

$$\mathbf{v} = \lambda\mathscr{H}'\mathbf{v} \tag{6.151}$$

does not have an eigenvalue $\lambda = 1$, then $\mathbf{w}$ is an isolated fixed point of index $(-1)^\alpha$, with α the sum of the multiplicities of the eigenvalues of eqn (6.151) which lie in [0, 1]. If $\lambda = 1$ is not an eigenvalue of eqn (6.151) the number of steady states is odd, for the indices can only be ± 1 and the index $\mathbf{I} - \mathscr{H}$ can be shown to be $+1$. In particular if there are no eigenvalues of eqn (6.151) in [0, 1] then the solution is unique.

In the case of a spherical region Ω, Gavalas uses Poincaré's inequality

$$\int_0^1 \rho^2\left(\frac{\mathrm{d}\mathbf{v}}{\mathrm{d}\rho}, \frac{\mathrm{d}\mathbf{v}}{\mathrm{d}\rho}\right)\mathrm{d}\rho \geqslant 6\int_0^1 \rho^2(\mathbf{v}, \mathbf{v})\,\mathrm{d}\rho \tag{6.152}$$

(where $(.,.)$ is the scalar product and $\mathbf{v}$ any vector function vanishing at $\rho = 1$) to get a bound on the size of the particle. In our notation we observe that eqn (6.151) is equivalent to

$$\frac{1}{\rho^2}\frac{\mathrm{d}}{\mathrm{d}\rho}\left(\rho^2\frac{\mathrm{d}\mathbf{v}}{\mathrm{d}\rho}\right) + \lambda\mathbf{P}'\mathbf{v} = 0, \tag{6.153}$$

where

$$(\mathbf{P}')_{jk} = \phi_j^2(\partial P_j/\partial w_k). \tag{6.154}$$

Forming the scalar product of eqn (6.153) with $\rho^2\mathbf{v}$ and integrating over the sphere gives

$$\lambda = \left[\int_0^1 \rho^2\left(\frac{\mathrm{d}\mathbf{v}}{\mathrm{d}\rho}, \frac{\mathrm{d}\mathbf{v}}{\mathrm{d}\rho}\right)\mathrm{d}\rho\right] \Big/ \left[\int_0^1 \rho^2(\mathbf{v}, \mathbf{P}'\,\mathbf{v})\,\mathrm{d}\rho\right]. \tag{6.155}$$

if χ_M is the largest eigenvalue of the symmetric matrix $\mathbf{\Pi} = \frac{1}{2}\{\mathbf{P}' + (\mathbf{P}')^T\}$ then the denominator is less than $\chi_M(\mathbf{v}, \mathbf{v})$. Using Poincaré's inequality for the numerator, and this bound for the denominator, shows that $\lambda > 1$, and hence uniqueness is guaranteed, if

$$\chi_M < 6. \tag{6.156}$$

Comparison with the single reaction case shows that this condition is conservative, but, since χ_M is small when all the ϕ_j are small, it shows that uniqueness is obtained for sufficiently small particles. It also shows that if $\mathbf{\Pi}$ is negative definite (as is the case with endothermic reactions) then uniqueness is assured for all values of the Thiele moduli.

Lumping techniques have been used in various ways, though more in connection with stability studies than in pursuing the question of uniqueness. Hlaváček, Kubíček and Višňák (1972) have considered the behaviour of the stirred tank with exothermic sequential reactions A → B → C and studied the range of parameter space over which there are one, three or five solutions. Such a study is of interest in itself and would be of value in the present context if the lumping procedure could be shown to be reliable. Amundson and Kuo (1967*a*, *b*, *c*) studied monomolecular reaction systems through an increasingly complex series of models. In the first paper all resistances were lumped at the surface and the stirred tank equations therefore obtained. In the second paper the pellet is held to be isothermal at a temperature determined by the Biot number for heat transfer and the concentrations are governed by the diffusion-reaction equation. The computation of the steady state is discussed and an example of five steady states given, but no general observations on uniqueness are made, since the emphasis is on questions of stability. In the third paper both the concentrations and the temperature are taken to be functions of position.

In general the conclusions that can be drawn for multiple reactions are more vague and the estimates cruder than those that can be made for a single reaction.

6.6. Cognate problems

We have noticed in Section 4.3 that the exponential approximation gives some insight into the more difficult problem of the exothermic reaction and have earlier in Section 3.4 had a form of the Emden–Fowler equation under consideration. Therefore, though its bearing on diffusion and reaction problems is indirect, it will be well to comment further on a form of Emden equation studied by Joseph and Lundgren (1972). In any case their results are of such unusual elegance and completeness that it would be unpardonable to pass them by altogether. They consider the equation

$$(\rho^q w')' + \lambda \rho^q P(w) = 0, \qquad \lambda > 0, \tag{6.157}$$

in the unit ball of R^{q+1} with boundary conditions

$$w'(0) = w(1) = 0, \tag{6.158}$$

and

$$P(w) = (1+\alpha w)^{\beta}, \qquad \alpha\beta > 0; \tag{6.159}$$

or its limit as $\beta = 1/\alpha \to \infty$, namely

$$P(w) = \mathrm{e}^{w}. \tag{6.160}$$

Thus for $q = 0, 1, 2$, $\lambda = \delta$ and the form (6.160) we have the Frank-Kamenetskii equation considered in Section 4.3.2. Unfortunately other values of q do not have physical meaning in our context, though they do in astrophysics (see e.g. Chandrasekhar, 1939), but either form of equation can provide an interesting approximation to the highly temperature-dependent heat generation functions $P(w)$ which we have encountered. It should be noted that the type of solution we have encountered with $w(\rho)$ identically constant in $0 \leqslant \rho \leqslant \omega$ plays no role in much of the analysis of the Emden equation. When this type of solution is ignored there is often, as we have seen in Section 4.3.2, a limit λ^* to the range of positive solutions of eqns (6.157) and (6.158). In fact if μ_1 and $u_1(\rho)$ are the least eigenvalue and corresponding eigenfunction of

$$(\rho^q u')' + \mu\rho^q u = 0, \qquad u'(0) = u(1) = 0,$$

then

$$\lambda = \mu_1 \left\{ \int_0^1 \rho^q u_1 w \,\mathrm{d}\rho \right\} \bigg/ \left\{ \int_0^1 \rho^q u_1 P(w) \,\mathrm{d}\rho \right\}.$$

Hence if $Q(w) = P(w)/w$ is bounded below away from zero, the right-hand side of this equation is bounded above, and for sufficiently large λ there can be no solution.

Since the maximum of $w(\rho) = \omega(\lambda)$ occurs at $\rho = 0$, it will be convenient to regard it, rather than λ, as preassigned and solve the equation (6.157) subject to

$$w(0) = \omega, \qquad w'(0) = 0$$

determining the value of λ so that $w(1) = 0$. The reason for this is that, as we have experienced so many times, $\lambda(\omega)$ is single valued though $\omega(\lambda)$ is not. Some calculated results of Joseph and Lundgren are shown in Fig. 6.22. We have encountered curves similar to those for $q = 1$ and 2, the latter showing the feature that at a critical value $\lambda = \lambda_c$ there are infinitely many solutions. What is most striking is that when $q \geqslant 9$ the solution is unique for $\lambda < \lambda^*$, and the high degree of multiplicity experienced in $1 < q < 9$ is lost in a larger number of dimensions.

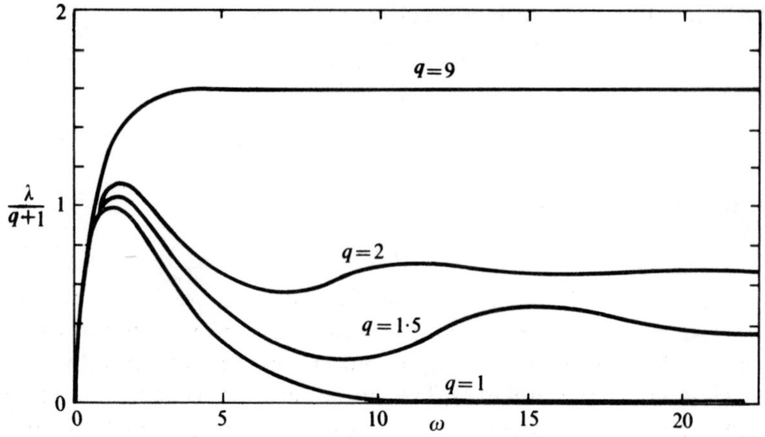

FIG. 6.22. The parameter λ as a function of ω for the solution of the Emden equation with various q. (After Joseph and Lundgren.)

The results of Joseph and Lundgren may be summarized in two theorems.

THEOREM I. If $\lambda > 0$ and $q > 1$ are given, the positive solutions of eqns (6.157) and (6.158) with the function (6.159) have the following properties:

(a) when $0 \leqslant \beta < 1$ the solutions are unique;
(b) when $\beta < 0$ or $\beta > 1$, there exists a positive λ^* such that the solution does not exist for $\lambda > \lambda^*$; a unique solution exists if $\lambda = \lambda^* > \lambda_c$ (defined below);
(c) when $1 < \beta \leqslant (q+3)/(q-1)$ and $\lambda < \lambda^*$ there are just two solutions;
(d) when $\beta < 0$ or $\beta > (q+3)/(q-1)$ and $q < \phi(\beta)$ and $\lambda = \lambda_c < \lambda^*$ there are a countably infinite number of solutions; an arbitrarily large but finite number of solutions can be obtained for λ sufficiently close to λ_c;
(e) when $\beta < 0$ or $\beta > (q+3)/(q-1)$ and $q \geqslant \phi(\beta)$ and $\lambda < \lambda^* = \lambda_c$ there is a unique solution.

In the above statements

$$\lambda_c = \frac{2(q-1)(\beta-1)-4}{\alpha(\beta-1)^2} \tag{6.161}$$

and

$$\phi(\beta) = \left\{1+2\sqrt{\left(\frac{\beta}{\beta-1}\right)}\right\}^2. \tag{6.162}$$

THEOREM II. If $\lambda > 0$ and $q > 1$ are given, the positive solutions of eqns (6.157) and (6.158) with the function (6.160) have the following properties:

(a) there exists a value λ^* such that there are no solutions for $\lambda > \lambda^*$; when $q \geqslant 9$ $\lambda^* = \lambda_c = 2(q-1)$;

(b) when $q \geqslant 9$ there is a unique solution for each $\lambda < \lambda^*$;
(c) when $q < 9$ and $\lambda = \lambda^*$ there is a unique solution;
(d) when $q < 9$ and $\lambda = \lambda_c = 2(q-1)$ there are a countably infinite number of solutions;
(e) when $q < 9$ an arbitrarily large number of solutions can be obtained by taking $|\lambda - 2q + 2| \neq 0$ sufficiently small.

These are most remarkable results and, while it would be out of place to derive them in full, since the reader can go to the original paper of Joseph and Lundgren, it is worth describing the approach that these authors take. First we note that the uniqueness of the solution for $0 \leqslant \beta \leqslant 1$ can be established from the theorems of Cohen and Keller. Let Ω be the first value of w for which $P(w)$ is infinite, i.e. for $\beta > 0, \Omega = \infty$, but for $\beta < 0, \Omega = 1/|\alpha|$, since the condition $\alpha\beta > 0$ implies that $\alpha < 0$ and $P(w) = (1 - |\alpha|w)^{-|\beta|}$. Then a limiting singular solution $(\tilde{w}(\rho), \lambda)$ is the limit of a regular solution $(w(\rho, \omega), \lambda(\omega))$ as $\omega \to \Omega$ which itself satisfies the equation everywhere except at the origin. The possible forms of singular solutions are

$$w_c(\rho) = (\rho^{-\tau} - 1)/\alpha, \tag{6.163}$$

where

$$\lambda_c = \{2(q-1)(\beta-1) - 4\}/\alpha(\beta-1)^2 \tag{6.164}$$

and

$$\tau = 2/(\beta - 1). \tag{6.165}$$

These are called singular similarity solutions and will be singular limiting solutions if $w(\rho, \omega) \to w_c(\rho)$ and $\lambda(\omega) \to \lambda_c$ as $\omega \to \Omega$. The possibility of establishing this rests on a similarity transformation:

$$v = \frac{1+\alpha w}{1+\alpha\omega}, \qquad \sigma = \rho\left(\frac{\lambda}{\lambda_c}\right)^{\frac{1}{2}}(1+\alpha\omega)^{\frac{1}{2}(\beta-1)} = \sigma_c\rho. \tag{6.166}$$

The equation for $v(\sigma)$ is

$$\frac{1}{\sigma^q}\frac{\mathrm{d}}{\mathrm{d}\sigma}\left(\sigma^q \frac{\mathrm{d}v}{\mathrm{d}\sigma}\right) + \lambda_c \alpha v^\beta = 0, \tag{6.167}$$

with

$$v(0) = 1, \qquad v'(0) = 0 \tag{6.168}$$

and

$$v(\sigma_c) = (1+\alpha\omega)^{-1}. \tag{6.169}$$

The solution of eqns (6.167) and (6.168) can be obtained by a single forward integration for any pair of values of β and q. From this solution we can work back to the original problem, for, given α and ω, the point on the solution satisfying eqn (6.169) may be found and then the definition of σ_c in eqn (6.166) gives $\lambda(\omega)$. Moreover eqns (6.167) and (6.168) have the property that if $v(\sigma)$ is a solution so also is $\kappa^{-2/(\beta-1)}v(\sigma/\kappa)$ for any κ.

It can now be seen that, when $\beta > 1$ or $\beta < 0$, $w(\rho, \omega)$ approaches $w_c(\rho)$ if it is derived from a solution of eqns (6.167) and (6.168) which is positive and tends to $\sigma^{-\tau}$ as $\sigma \to \infty$. This follows if we write eqn (6.166) as

$$\begin{aligned}1+\alpha w(\rho,\omega) &= (1+\alpha\omega)v\left[\left(\frac{\lambda}{\lambda_c}\right)^{\frac{1}{2}}(1+\alpha\omega)^{\frac{1}{2}(\beta-1)}\rho\right]\to \\ &\to(1+\alpha\omega)\left(\frac{\lambda_c}{\lambda}\right)^{\tau/2}(1+\alpha\omega)^{-1}\rho^{-\tau},\end{aligned} \tag{6.170}$$

since as $\omega \to \Omega$ the argument of v goes to infinity for any fixed ρ. Thus $w(\rho, \omega) \to w_c(\rho)$ if $\lambda \to \lambda_c$ as $\omega \to \Omega$. The limit does not hold if there is a finite point Σ such that in the integration of eqns (6.167) and (6.168) $v(\Sigma) = 0$ for $\beta > 1$ or is infinite for $\beta < 0$. For in that case λ would have to tend to zero as $\omega \to \Omega$ to keep the argument of v finite.

If new variables

$$r(\zeta) = \sigma^\tau v(\sigma), \qquad \zeta = \ln(1/\sigma), \qquad s = \mathrm{d}r/\mathrm{d}\zeta \tag{6.171}$$

are introduced, eqn (6.167) becomes

$$s\frac{\mathrm{d}s}{\mathrm{d}r}-(q-1-2\tau)s+\tau(q-1-\tau)(r^\beta-r) = 0, \tag{6.172}$$

with the initial conditions (6.163) replaced by

$$r \to \sigma^\tau, \qquad s \to -\tau\sigma^\tau \qquad \text{as } \zeta \to \infty. \tag{6.173}$$

It is the study of the phase plane of r and s which reveals the disposition of the solutions and, though we do not go into detail, it will be well to understand why this is the case. The original boundary value problem, (6.157) and (6.158), was transformed into an initial value problem, (6.167) and (6.168). This has now been transformed into a first-order equation, (6.172). Moreover ω does not appear in eqn (6.172) so that, by considering a point $r, s(r)$ on a solution curve of (6.172) for given r, each value of s gives a unique σ. Let $(r, s(r))$ be a point on the solution curve of eqn (6.172) for which $\sigma = \sigma_c$, and so $v = (1+\alpha\omega)^{-1}$. Then $r = \sigma_c^\tau v(\sigma)$ and

$$\lambda = \lambda_c r^{\beta-1}$$

is determined. The number of solutions for a given value of r (i.e. of λ) is the number of times the solution curve in the (r, s)-plane crosses the line of given

$r = \lambda_c$. The extreme value of r for which there is a crossing gives the value of λ^*. If the critical point $r = 1, s = 0$ is a focus then the solution curve comes into it in the form of a spiral and there are an infinite number of solutions for $\lambda = \lambda_c$. For a value of r very close to 1 there are a finite but large number of crossings. Thus a thorough study of the possible aspects of the phase plane leads to the conclusions stated in the summarizing theorems above.

This example has been treated at some length for the sake of the insight that it gives into the question of multiplicity. It is not that the heat generation functions correspond very faithfully with any such term which would arise with diffusion, but rather it shows that it is the steeply rising part of the heat generation function which is the important factor in accounting for the high multiplicity observed in, for example, the irreversible, exothermic reaction. It is because this part of the heat generation curve is increasingly steep and extended for large values of β and γ that the number of solutions can become large.

The tubular reactor is of course a closer analogue to the catalyst particle in both a physical and chemical sense. It differs mathematically in having convective derivative terms and in a general case would be represented by such equations as

$$\frac{\partial u}{\partial \tau} = \nabla^2 u = \mathbf{V} \cdot \nabla u - \phi^2 R(u, v),$$

$$\mathscr{L}\frac{\partial v}{\partial \tau} = \nabla^2 v - \mathbf{V} \cdot \nabla v + \beta\phi^2 R(u, v), \qquad \text{in } \Omega$$

with

$$\frac{\partial u}{\partial \nu} = \frac{\partial v}{\partial \nu} = 0$$

on $\partial\Omega_0$, those parts of $\partial\Omega$ which are impervious or through which the products leave, and

$$\frac{1}{\nu}\frac{\partial u}{\partial \nu} + u = 1, \qquad \frac{1}{\mu}\frac{\partial v}{\partial \nu} + v = 1$$

on $\partial\Omega_1$, the part of $\partial\Omega$ which serves as an inlet. Here $\mathbf{V}$ is the dimensionless velocity field within the reactor and it is this which distinguishes it from the catalyst particle. Some references are given in the bibliographical comments.

Additional bibliographical comments

6.1. For papers on the tubular reactor see Amundson and Raymond (1964), Amundson (1965), Amundson and Luss (1967*a*). Amundson in papers with Luss (1967*c*) and Varma (1972*b*), has considered both problems together. For a discussion of the reactor as a whole see Hlaváček (1970*a*, *b*), Hlaváček and Hofmann (1970*a*, *b*, *c*), Wicke (1961),

Wicke and Hugo (1968*a*), Wicke, Padberg and Arens (1968), Wicke and Fieguth (1971). Other studies of the slab have been made by Slinko, Beskov and Zelenyak (1966).

6.1.1. Hlaváček and Kubíček (1970*c*) have also considered the isothermal pellet model. They present regions of uniqueness and multiplicity in the ϕ, $(\beta\gamma/\mu)$-plane for constant γ and ν. Simchen (1964) uses Semenov's assumptions to get the equation $a(T-T_0) = \exp(-E/RT)$ for the steady state temperature in terms of the wall temperature T_0. He presents graphs and tables of the possible solutions. Østergaard (1963) considered the Dirichlet problem and showed the region of uniqueness by means of a graph of η versus γ for constant (β/γ) for both first and zero order reactions.

A recent paper by Stewart, Guertin and Sørensen (1973) gives bifurcation values for the finite cylinder of radius a and length $2l$. Values are listed for a modulus $\Lambda = \phi/(2+a/l)$ reduced for shape but not kinetics. Their paper lists Λ_b and Λ^b for $\gamma = 20$, $\beta = 0{\cdot}3$, $0{\cdot}4$, $0{\cdot}5$; $\gamma = 30$, $\beta = 0{\cdot}2\ (0{\cdot}1)\ 0{\cdot}5$; $\gamma = 40$, $\beta = 0{\cdot}15$, $0{\cdot}2$, $0{\cdot}3$ and for $a/l = 0$, $0{\cdot}25$, $0{\cdot}5$, 1, ∞. They also use the approximate method of the type devised by Villadsen and Stewart (1969) and discussed below in Section 6.4.2.

6.2. For various existence theorems the papers of Amann (1971, 1972*a*,*b*, 1973) are very helpful as are the papers of Cohen (1967*a*, *b*, 1972), of H. B. Keller (1968, 1969*a*, *b*, 1972) and their joint paper (1967). For an application of maximal and minimal solutions see Amundson and Varma (1973).

6.3. The proof of the sufficient condition $Q'(w) < 0$ is an adaptation of that in Luss (1971); see also Luss (1968). Similar criteria have been used by others, e.g. Joseph and Sparrow (1970), Cohen and Laetsch (1970).

6.4. The first results of this kind were obtained by Gavalas using a method based on Green's function. Since these are mentioned in Section 6.5 for multiple reactions they are not described in detail for the single reaction. Dente, Biardi and Ranzi (1969*b*) have used a Green's function iteration on the equation for $(\partial u/\partial\phi)$ as a means of finding the point of vertical tangency on the (η, ϕ)-curve. They give curves for the region of multiplicity in the (β, ϕ)-plane for a sphere with $\gamma = 20, 30, 40$ and $p = 0{\cdot}5, 1, 2$. They show also the effectiveness factor curves for $p = 0{\cdot}5$, $\gamma = 30$, $\beta = 0{\cdot}2, 0{\cdot}4, 0{\cdot}6, 0{\cdot}8$. For the last there is a region of five steady states, for which they show typical profiles and the region of the (β, ϕ)-plane.

6.4.1. Dente and Biardi (1967) use a one point approximation and compare their results with data from Weisz and Hicks (1972). They show the region of multiplicity in the (β, ϕ)-plane for $p = 1$; $q = 0, 1, 2$; $\gamma = 20, 30, 40$.

6.5. I have not attempted to give a condensed introduction to the notion of an index on which Gavalas's work relies. For this see the appendix of Gavalas's tract (1968) or the books of Krasnosel'skii (1964) and Cronin (1964).

6.6. Other problems with nonlinear heat source terms have been considered by Joseph (1964, 1965, 1966) and Joseph and Sparrow (1970), Regirer (1958) and Kaganov (1963).

Many papers link together the analysis of catalyst particle and tubular reaction: see for example, Amundson and Luss (1967*c*), Hlaváček (1970*b*); Luss (1971 and 1972); Ray (1972*a*); Amundson and Varma (1972*a*). The literature on the tubular reactor is too vast to be surveyed here; to the references given above under 7.1 may be added Amundson and Markus (1968), Cohen (1971), Cohen and Laetsch (1970), Cohen (1973).

NOMENCLATURE

See also General Nomenclature for commonly used symbols. Highly ephemeral nomenclature is not included.

E	maximum value defined by eqn (6.115)
$\mathcal{G}$	Green's function (Section 6.5)
g	equation for surface value $\tilde{w}$ (6.55)
$\mathcal{H}$	nonlinear operator on space of $\mathbf{w}$ (Section 6.5)
I_{ji}, I_{jo}	integrals defined in eqn (6.104)
M_j	integrands in I_j, eqn (6.103)
$P(w)$	$R(1-w, 1+\beta w)$, reaction rate for Dirichlet problem
$P(w;\tilde{w})$	$R(1-w-\tilde{w}, 1+\beta w+\alpha\beta\tilde{w})$, reaction rate for Robin problem
$\mathbf{P}'$	Jacobian matrix of rate expressions (Section 6.5)
$Q(w)$	$P(w)/w$
$Q(w;\tilde{w})$	$P(w;\tilde{w})/w$
r, s	phase plane variables in Section 6.6
U	maximum value defined in eqn (6.90)
$\mathbf{w}$	vector of concentrations and temperature, Section 6.5
$w_c(\rho)$	singular solution in Section 6.6
$\check{w}, \hat{w}$	minimal and maximal solutions
w_*, w^*, W_*, W^*	particular values of w defined in Fig. 6.14
$w', \tilde{w}'$	transformed w defined by eqn (6.68)
α, β	parameters in the Emden equation, Section 6.6
α	ν/μ
$\beta'\ \gamma'\ \tilde{\beta}', \tilde{\gamma}'$	transformed parameters defined by eqn (6.68)
δ	$\beta\gamma$
Γ_n	regions of maximum multiplicity in (β, γ)-space
λ	parameter in Section 6.6
λ_c	critical value for infinite number of solutions, Section 6.6
λ^*	maximum value for any solutions, Section 6.6
λ_1^2	least eigenvalue of linearized problem
$\mathbf{\Pi}$	symmetrized form of $\mathbf{P}'$
$\rho_{(1)}$	collocation point
σ, τ	variable and parameter in Section 6.6
$\phi(\beta)$	function defined by eqn (6.162)
ϕ_b, ϕ^b	bifurcation values of Thiele modulus

Φ_b, Φ^b	bifurcation values of normalized Thiele modulus
ϕ_t	trifurcation value of Thiele modulus
ψ_*, ψ^*	estimates of ϕ_b and ϕ^b
ψ^2	equivalent of Thiele modulus for stirred tank
χ_M	greatest eigenvalue of $\mathbf{\Pi}$
$\omega(\phi)$	maximum value of $w(\boldsymbol{\rho})$
Ω	limiting value of ω in Section 6.6

7

THE STABILITY OF THE STEADY STATE

7.1. Introduction

THE concepts of uniqueness and stability must be carefully distinguished. It is a property of the whole equation to have, or not to have a unique solution as the case may be. It is a property of any one solution to be stable or unstable. Questions of uniqueness or multiplicity of the steady state can be resolved by the study of the steady state equations, whereas questions of stability require the consideration of the transient equations. However the steady state equations do permit certain limited, and usually negative, statements to be made about stability. For example, when there are multiple solutions of the steady state equation not all the solutions can be stable. Indeed Gavalas (1968) in showing that, except under exceptional circumstances, there were $2m+1$ steady states, showed also that at least m of them would be unstable. Moreover any solution with index -1 must be unstable.

We have seen that the general form of the equations is

$$\frac{\partial u_i}{\partial \tau} = \Delta_i \nabla^2 u_i + \sum_{j=1}^{R} \alpha_{ji} \phi_j^2 R_j(u, v). \tag{7.1}$$

$$\mathscr{L}\frac{\partial v}{\partial \tau} = \nabla^2 v + \sum_{j=1}^{R} \beta_j \phi_j^2 R_j(u, v) \tag{7.2}$$

in Ω, where for simplicity we have chosen the case of independent and constant diffusion coefficients. These equations are subject to boundary conditions on $\partial\Omega$,

$$\frac{\Delta_i}{\nu_i}\left(\frac{\partial u_i}{\partial \nu}\right)_{\mathrm{s}} + u_{i\mathrm{s}} = u_{i\mathrm{f}}, \qquad \frac{1}{\mu}\left(\frac{\partial v}{\partial \nu}\right)_{\mathrm{s}} + v_{\mathrm{s}} = 1, \tag{7.3}$$

and initial values in $\partial\Omega$

$$u_i(\boldsymbol{\rho}, 0) = u_{i0}(\boldsymbol{\rho}), \qquad v(\boldsymbol{\rho}, 0) = v_0(\boldsymbol{\rho}). \tag{7.4}$$

There are three classes of parameters or parametric function involved in these equations:

(i) the stoicheiometric coefficients, α_{ji};

(ii)*a* the diffusion coefficient ratios, Δ_i; the concentrations $u_{i\mathrm{f}}$;

(ii)*b* the Thiele moduli, ϕ_j; the Prater temperatures, β_j; the Biot numbers, ν_i, μ; at least one other parameter in each R_j, such as a reaction order or activation energy;

(iii) the Lewis number, $\mathscr{L}$; the functions u_{i0} and v_0.

The first class has been separated out because it is independent both of the diffusion process and of the kinetics of the reaction. The second has been subdivided only for the convenience of the fact that we have been accustomed to keeping those in (ii)*a* fixed and considering the influence of those in (ii)*b*. If multicomponent diffusion were being considered we would include the full sets of multicomponent diffusion coefficients, Δ_{ik}; if the diffusion coefficients and conductivity were variable, these terms would be within the divergence operation and further parameters expressing this dependency would be needed. It is on all the quantities in classes (i) and (ii) that the steady states depend and it is only a matter of convenience that we have concentrated on understanding the influence of those in class (ii)*b*. The parameters in class (iii) govern the transient behaviour, for it will be found that the stability of the steady state depends on $\mathscr{L}$ and the choice of steady state that is finally reached is governed by the initial distributions of concentration and temperature. In addition there may be invariant sets of the state space which are solutions of the transient equations but are not steady states.

On physical grounds we would hope that for each instant the concentrations, $u_i(\boldsymbol{\rho}, \tau)$, and the temperature, $v(\boldsymbol{\rho}, \tau)$, would lie in the space $C_+(\Omega)$ of continuous, positive functions defined on Ω. It is of course the business of an existence theorem to establish the conditions under which this can be asserted. In whatever space the S concentrations and the temperature can be shown to lie, the state of the system is represented by a 'point' in the $(S+1)$-fold Cartesian product of this space. A point in this so-called state space, $\mathscr{S}$, is a vector of functions

$$\mathbf{w} = \mathbf{w}(\tau) = \{u_1(\boldsymbol{\rho}, \tau), \dots, u_s(\boldsymbol{\rho}, \tau), v(\boldsymbol{\rho}, \tau)\}.$$

An invariant set $\mathscr{I}$ is a subset of $\mathscr{S}$ such that if

$$\mathbf{w}_0 = \{u_{10}(\boldsymbol{\rho}), \dots, u_{s0}(\boldsymbol{\rho}), v_0(\boldsymbol{\rho})\} \in \mathscr{I},$$

then the solution of eqns (7.1–4) remains in $\mathscr{I}$ for all τ and moreover this is not true of any subset of $\mathscr{I}$. The steady states are invariant sets in $\mathscr{S}$ since if the initial distributions are so given they will remain the same for all time. But there can also be invariant sets that are limit cycles in $\mathscr{S}$ and correspond to continually oscillating functions of position in Ω. Even more complicated behaviour can be imagined but this will suffice. Associated with each invariant set $\mathscr{I}_p$ is an open subset $\mathscr{S}_p$ of $\mathscr{S}$ within which it lies and such that, if $\mathbf{w}_0 \in \mathscr{S}_p$ then $\mathbf{w}(\tau)$ tends to $\mathscr{I}_p$: $\mathscr{S}_p$ is called the region of attraction of $\mathscr{I}_p$. An invariant set is stable if its region of attraction $\mathscr{S}_p$ is a true neighbourhood of it in the space $\mathscr{S}$, for then any sufficiently small perturbation will die away.

A full description of the solution of eqns (7.1–4) would answer the following three questions:

(a) given the parameters in classes (i) and (ii) what are the invariant sets?

(b) are they stable or unstable?

(c) what is the region of attraction associated with each of the invariant sets?

As usual it is rarely possible to answer these questions as fully as would be desirable. The first has engaged us up to this point and the second and third will be our concern in this chapter. The third is the most difficult since describing a region of function space is far from easy. At best one hopes to get sections of it by some obvious class of functions. For example Satterfield, Roberts, and Hartman (1967) considered the Langmuir–Hinshelwood kinetics which are capable of giving multiple steady states and show that among all constant initial distributions of the reactant A those with Kp_{AO} below a certain critical value lead to one stable steady state, whereas those with Kp_{AO} greater than this value lead to another. Luss and Lee (1968) also have found sections by initially constant functions.

In this chapter we shall describe some of the methods that have been used to discuss the stability of steady states: the limit cycle as an invariant set is reserved for the next chapter. We shall take up methods based on the maximum principle or comparison theorems, on linearization and on Lyapounov's direct method. Following a discussion of approximate methods we shall look at particular systems, in particular the catalytic wire and the unsymmetrical solutions.

To conclude this introduction we should give a more formal definition of stability. Let $\|\,.\,\|$ denote a suitable norm in $\mathscr{S}$, then a steady state of the system (7.1–4) is a solution $\mathbf{w}_e(\boldsymbol{\rho})$ of (7.1–3) such that if $\mathbf{w}_0(\boldsymbol{\rho}) = \mathbf{w}_e(\boldsymbol{\rho})$ then $\|\mathbf{w}(\boldsymbol{\rho}, \tau) - \mathbf{w}_e(\boldsymbol{\rho})\| = 0$ for all $\tau \geqslant 0$. The steady state $\mathbf{w}_e(\boldsymbol{\rho})$ is said to be stable if given any $\varepsilon > 0$ there exists a δ such that $\|\mathbf{w}_0 - \mathbf{w}_e\| < \delta$ implies $\|\mathbf{w} - \mathbf{w}_e\| < \varepsilon$, $\tau \geqslant 0$. The steady state is asymptotically stable if it is stable and $\|\mathbf{w} - \mathbf{w}_e\| \to 0$ as $\tau \to \infty$. If in addition there are positive constants M and μ such that $\|\mathbf{w} - \mathbf{w}_e\| \leqslant M\,\mathrm{e}^{-\mu\tau}\|\mathbf{w}_0 - \mathbf{w}_e\|$ for all $\tau \geqslant 0$ then the steady state is called exponentially asymptotically stable. If δ can be so large that any initial state $\mathbf{w}_0(\boldsymbol{\rho})$ is contained in the region of attraction of the steady state $\mathbf{w}_e(\boldsymbol{\rho})$, then the latter is said to be globally stable. Only a unique steady state can be globally stable.

It is often desirable to work with the difference between the solution and the steady state, which will be distinguished by an asterisk. Thus

$$\left.\begin{aligned} u_i^*(\boldsymbol{\rho}, \tau) &= u_i(\boldsymbol{\rho}, \tau) - u_{ie}(\boldsymbol{\rho}) \\ v^*(\boldsymbol{\rho}, \tau) &= v(\boldsymbol{\rho}, \tau) - v_e(\boldsymbol{\rho}) \end{aligned}\right\} \tag{7.5}$$

satisfy

$$\frac{\partial u_i^*}{\partial \tau} = \Delta_i \nabla^2 u_i^* + \sum_{j=1}^{R} \alpha_{ji} \phi_j^2 R_j^*(u^*, v^*, \boldsymbol{\rho}), \tag{7.6}$$

$$\frac{\partial v^*}{\partial \tau} = \nabla^2 v^* + \sum_{j=1}^{R} \beta_j \phi_j^2 R_j^*(u^*, v^*, \boldsymbol{\rho}) \tag{7.7}$$

in Ω, with

$$\frac{\Delta_i}{\nu_i}\left(\frac{\partial u_i^*}{\partial \nu}\right)_{\mathrm{s}} + (u_i^*)_{\mathrm{s}} = 0, \qquad \frac{1}{\mu}\left(\frac{\partial v^*}{\partial \nu}\right) + v^* = 0 \tag{7.8}$$

on $\partial\Omega$, where

$$R_j^*(u^*, v^*, \boldsymbol{\rho}) = R_j(u, v) - R_j(u_{\mathrm{e}}, v_{\mathrm{e}}). \tag{7.9}$$

For a single equation, such as the Dirichlet problem for an isothermal irreversible reaction some theorems of Pao and Vogt (1969) can be applied immediately. Thus let

$$\frac{\partial u}{\partial \tau} = \frac{\partial^2 u}{\partial \rho^2} - \phi^2 R(u) \tag{7.10}$$

with $u = 1$ on $\rho = \pm 1$ be the single defining equation. The deviation u^* satisfies

$$\frac{\partial u^*}{\partial \tau} = \frac{\partial^2 u^*}{\partial \rho^2} - \phi^2 R^*(u^*, \rho), \tag{7.11}$$

where

$$R^*(u^*, \rho) = R(u_{\mathrm{e}}(\rho) + u^*) - R(u_{\mathrm{e}}(\rho)) \tag{7.12}$$

and

$$u^* = 0 \quad \text{on } \partial\Omega. \tag{7.13}$$

Then a theorem of Pao and Vogt states that the null solution, $u^* \equiv 0$, is exponentially asymptotically stable if

$$\phi^2 \int_0^1 \{R^*(u_1^*, \rho) - R^*(u_2^*, \rho)\}(u_1^* - u_2^*)\,\mathrm{d}\rho < \pi^2 \int_0^1 (u_1^* - u_2^*)^2\,\mathrm{d}\rho. \tag{7.14}$$

For example, if

$$R(u) = \frac{(1+\kappa)^2 u}{(\kappa + u)^2}, \tag{7.15}$$

the integral on the left of eqn (7.14) is

$$\phi^2(1+\kappa)^2 \int_0^1 \frac{\kappa^2 - u_1 u_2}{(\kappa + u_1)^2(\kappa + u_2)^2}(u_i^* - u_2^*)^2\,\mathrm{d}\rho \leqslant \frac{\phi^2(1+\kappa)^2}{\kappa^2}\int_0^1 (u_i^* - u_2^*)^2\,\mathrm{d}\rho,$$

since the first factor in the integrand is not greater than 1; hence

$$\phi < \pi\kappa/(1+\kappa) \tag{7.16}$$

ensures stability. We observe that this condition also implies uniqueness so that we have a case of a unique stable steady state. The method of Pao and Vogt depends on providing certain contraction properties, so that we are not surprised to find that it leads to a condition on ϕ.

7.2. Sufficient conditions for instability

It is intrinsically easier to prove instability than it is to establish stability. For to establish stability we have to show that *every* sufficiently small perturbation dies away, whereas instability will be demonstrated if we can show that sufficiently small perturbations in only *one* direction away from the steady state will grow with time. This is illustrated by Jackson's geometric criterion for instability (Jackson (1973)) of symmetric solutions in symmetric bodies.

The pth-order irreversible reaction will be chosen to illustrate this and the argument is confined to the case of unit Lewis number and the Dirichlet boundary conditions. Then the equations

$$\frac{\partial u}{\partial \tau} = \frac{1}{\rho^q}\frac{\partial}{\partial \rho}\left(\rho^q \frac{\partial u}{\partial \tau}\right) - \phi^2 u^p \, e^{\gamma - \gamma/v} \tag{7.17}$$

$$\frac{\partial v}{\partial \tau} = \frac{1}{\rho^q}\frac{\partial}{\partial \rho}\left(\rho^q \frac{\partial v}{\partial \tau}\right) + \beta\phi^2 u^p \, e^{\gamma - \gamma/v} \tag{7.18}$$

subject to

$$\begin{gathered}\frac{\partial u}{\partial \rho} = \frac{\partial v}{\partial \rho} = 0, \quad \rho = 0, \\ u = v = 1, \quad \rho = 1,\end{gathered} \tag{7.19}$$

can be replaced by equations in v and the residual enthalpy,

$$z = \beta u + v. \tag{7.20}$$

These are

$$\frac{\partial v}{\partial \tau} = \frac{1}{\rho^q}\frac{\partial}{\partial \rho}\left(\rho^q \frac{\partial v}{\partial \rho}\right) + \beta\phi^2 F(v, z), \tag{7.21}$$

$$\frac{\partial z}{\partial \tau} = \frac{1}{\rho^q}\frac{\partial}{\partial \rho}\left(\rho^q \frac{\partial z}{\partial \rho}\right), \tag{7.22}$$

with

$$F(v, z) = \beta^{-p}(z - v)^p \, e^{\gamma - \gamma/v} \tag{7.23}$$

and

$$\frac{\partial v}{\partial \rho} = \frac{\partial z}{\partial \rho} = 0, \quad \rho = 0,$$
$$v = 1, \qquad z = 1+\beta, \qquad \rho = 1. \tag{7.24}$$

The initial distributions $u_0(\rho)$ and $v_0(\rho)$ will give initial distributions of v and z and so complete the statement of eqns (7.21–4). However, because we can choose to investigate a very special class of perturbations when we wish to prove instability, we shall insist that $z(\rho, \tau)$ never departs from its steady state value, the constant function $z(\rho) = 1+\beta$.

If $v_e(\rho)$ is the steady state we wish to investigate, it satisfies

$$0 = \frac{1}{\rho^q}\frac{\mathrm{d}}{\mathrm{d}\rho}\left(\rho^q \frac{\mathrm{d}v_e}{\mathrm{d}\rho}\right) + \beta\phi^2 F(v_e), \tag{7.25}$$

where

$$F(v) = \{(1+\beta-v)/\beta\}^p\, \mathrm{e}^{\gamma-\gamma/v} \tag{7.26}$$

and

$$v_e'(0) = 0, \qquad v_e(1) = 1. \tag{7.27}$$

The departure from steady state

$$v^*(\rho, \tau) = v(\rho, \tau) - v_e(\rho) \tag{7.28}$$

thus satisfies

$$\frac{\partial v^*}{\partial \tau} = \frac{1}{\rho^q}\frac{\partial}{\partial \rho}\left(\rho^q \frac{\partial v^*}{\partial \rho}\right) + \beta\phi^2\{F(v_e+v^*) - F(v_e)\}, \tag{7.29}$$

$$\frac{\partial v^*}{\partial \rho} = 0, \quad \rho = 0; \qquad v^* = 0, \quad \rho = 1. \tag{7.30}$$

If v^* is sufficiently small the difference in the last term of eqn (7.29) may be replaced by

$$-H(\rho)v^* = \beta\phi^2 F'(v_e(\rho))v^*, \tag{7.31}$$

so that eqn (7.29) becomes a linear equation in v^* with homogeneous boundary conditions. Its solution can be found by separation of variables and is

$$v^*(\rho, \tau) = \sum_{n=1}^{\infty} V_n(\rho)\exp(-\lambda_n \tau) \tag{7.32}$$

where

$$\frac{1}{\rho^q}\frac{\partial}{\partial\rho}\left(\rho^q\frac{\mathrm{d}V_n}{\mathrm{d}\rho}\right)-\{H(\rho)-\lambda_n\}V_n=0,$$
$$V_n'(0)=V_n(1)=0. \tag{7.33}$$

If we can prove that at least one of the eigenvalues of this equation is negative then $\|v^*\|$ will grow without bound however small $\|v_0-v_e\|$ may be and the steady state will be unstable.

The method Jackson uses to establish the condition for a negative eigenvalue is based on a comparison of Sturm. Such a technique was used first on reactor problems by Amundson (1965) using an example with $q = 0$. Jackson shows that this can be extended to $q = 1$ and 2. The result was first obtained by Villadsen and Michelsen (1972) but we shall follow Jackson's approach here. Consider

$$\frac{1}{\rho^q}\frac{\mathrm{d}}{\mathrm{d}\rho}\left(\rho^q\frac{\mathrm{d}U}{\mathrm{d}\rho}\right)-H(\rho)U=0, \tag{7.34}$$

subject to

$$U'(0)=0, \qquad U(0)=1. \tag{7.35}$$

This is to be compared with the equation

$$\frac{1}{\rho^q}\frac{\mathrm{d}}{\mathrm{d}\rho}\left(\rho^q\frac{\mathrm{d}V}{\mathrm{d}\rho}\right)-\{H(\rho)-\lambda\}V=0, \tag{7.36}$$

subject to

$$V'(0)=0, \qquad V(0)=1. \tag{7.37}$$

The Sturmian theorem states that if $\lambda < 0$ then the function $U(\rho)$ will have at least as many zeros in the interval (0, 1) as does $V(\rho)$, and in fact the ith zero of U will occur at a smaller value of ρ than the ith zero of V. Indeed, as Jackson shows, the position of the zero will move continuously as λ varies. Suppose that U has just one zero in (0, 1) then the situation will be as shown in Fig. 8.1 with $U(\rho) = 0$ at the point A. For $\lambda = 0$ the curve $V(\rho)$ coincides with $U(\rho)$ and it has its zero also at A, but as λ decreases from zero the curve $V(\rho)$ lies above that of $U(\rho)$ and the position of the zero moves outward, as at B. Hence it follows that, by decreasing λ sufficiently, a value will be found at which the zero just reaches $\rho = 1$. This is a negative eigenvalue of eqn (7.33). On the other hand, if the first zero of $U(\rho)$ is beyond $\rho = 1$, then this construction shows that there can be no negative eigenvalue of eqn (7.33). It is therefore a sufficient condition for instability that the solution of eqns (7.34) and (7.35) should have a zero in the interval (0, 1).

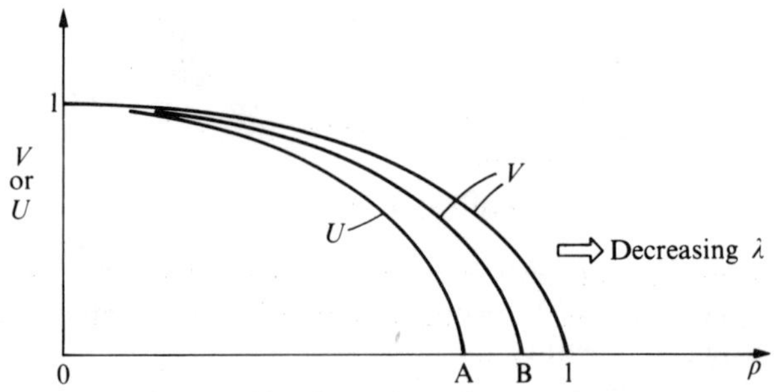

FIG. 7.1. The relative disposition of the solution of eqns (7.34) and (7.36).

We now wish to relate this criterion to the geometry of the (η, ϕ)-curve. Except at a bifurcation value of ϕ the solution $v_e(\rho)$ and the effectiveness factor are continuously differentiable functions of ϕ and

$$\dot{v}_e(\rho) = \frac{\partial}{\partial \phi} v_e(\rho) \tag{7.38}$$

satisfies

$$\frac{1}{\rho^q}\frac{\mathrm{d}}{\mathrm{d}\rho}\left(\rho^q \frac{\mathrm{d}\dot{v}_e}{\mathrm{d}\rho}\right) - H(\rho)\dot{v}_e = -2\beta\phi F(v_e). \tag{7.39}$$

But if we multiply eqn (7.39) by $\rho^q U(\rho)$, eqn (7.34) by $\rho^q \dot{v}_e$, and subtract, we have, on integrating over (0, 1),

$$\left(\frac{\mathrm{d}\dot{v}_e}{\mathrm{d}\rho}\right)_{\rho=1} = \dot{v}_e'(1) = -\frac{2\beta\phi}{U(1)}\int_0^1 \rho^q F(v_e(\rho))U(\rho)\,\mathrm{d}\rho. \tag{7.40}$$

On the other hand

$$\phi^2\eta = -\frac{q+1}{\beta} v_e'(1),$$

and so

$$\frac{\mathrm{d}}{\mathrm{d}\phi}(\phi^2\eta) = \frac{2(q+1)\phi}{U(1)}\int_0^1 \rho^q F(v_e(\rho))U(\rho)\,\mathrm{d}\rho \tag{7.41}$$

Now since $F(v) > 0$, a sufficient condition for instability will be

$$\frac{\mathrm{d}}{\mathrm{d}\phi}(\phi^2\eta) < 0, \tag{7.42}$$

for this can only be the case if there is at least one zero of $U(\rho)$ in $0 \leqslant \rho \leqslant 1$. If

$$W(\rho) = \{U(\rho)/U(1)\} - 1 \tag{7.43}$$

then

$$\frac{d\eta}{d\phi} = \frac{2(q+1)}{\phi} \int_0^1 \rho^q F(v_e(\rho)) W(\rho)\, d\rho. \tag{7.44}$$

We have seen that existence of non-trivial solutions of the derived equation is just the condition for a bifurcation value of ϕ. Moreover the equation for U is just the derived equation, so that whenever $H(\rho)$ is such that $U(1) = 0$ we have a bifurcation situation. The solution may be divided up into branches by the points at which $d\eta/d\phi$, and so also $d(\eta\phi^2)/d\phi$, becomes infinite. In Fig. 7.2 this is shown for an (η, ϕ)-curve having at most five steady states. The branches of the solution are DA, AB, BC, CD and DE, the first and last corresponding to the minimal and maximal solutions in the interval $\phi_b < \phi < \phi^b$. Now the stability can only change at a bifurcation point, for $v_e(\rho)$, and so also $H(\rho)$, depends continuously on ϕ. Hence a change from

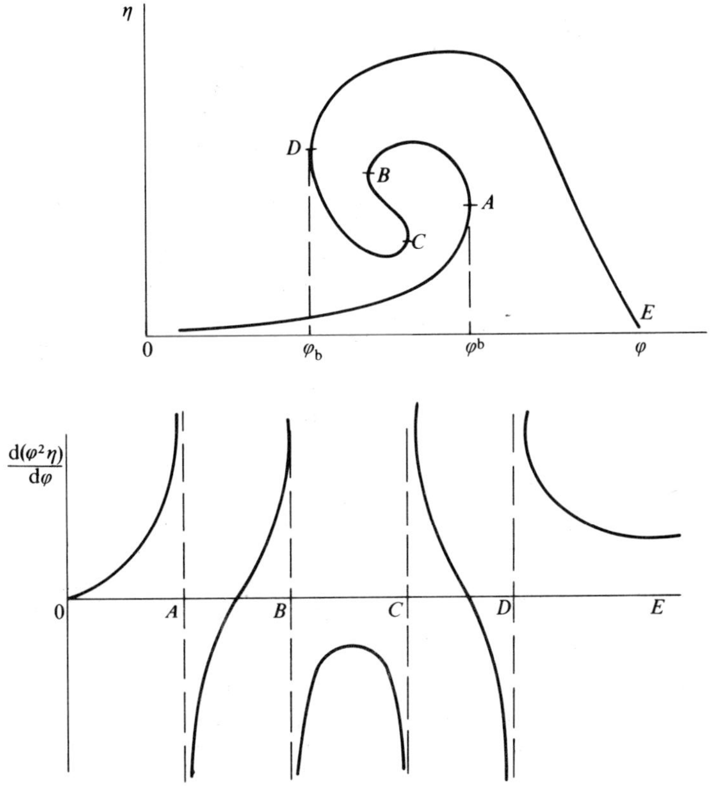

FIG. 7.2. Branches of the solution for exothermic reaction in the sphere with the variation of $d(\eta\phi^2)/d\phi$.

stability to instability, or vice versa, corresponds to the movement of the first zero of $U(\rho)$, defined by eqns (7.34) and (7.35), into or out of (0, 1), and this can only happen when the derived equation has a non-trivial solution. Thus all solutions on the same branch have the same stability character. But eqn (7.42) shows that any branch for which $d(\phi^2\eta)/d\phi$ is negatively infinite at an endpoint is unstable, and if we trace the sign of this derivative, as in the lower part of Fig. 7.2, we see that the branches AB, BC, and CD must give unstable steady states. We have yet to prove that those on OA or DE are stable, but certainly all solutions other than the maximal and minimal must be unstable. This accords with the result of Villadsen and Michelson (1972) who expressed in terms of the properties of the solution rather than of the effectiveness factor curve. It also agrees with Fujita's findings (1969) on the Gelfand equation.

It is instructive to notice that the proof carries over with minor changes to the special Robin problem with $\mu = \nu$. Since we know that the solution

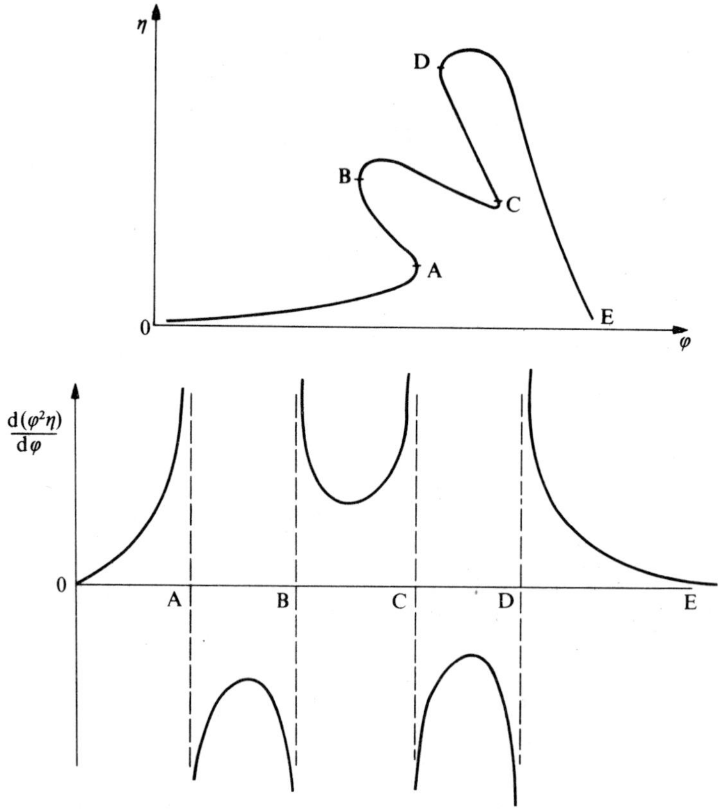

FIG. 7.3. Branches of the solution of the Robin problem with the variation of $d(\eta\phi^2)/d\phi$.

to the Robin problem may have a region of five steady states, as illustrated in Fig. 7.3, we may well ask whether only the maximal and minimal solutions can be stable. The variation of $d(\phi^2\eta)/d\phi$ is shown schematically in the lower part of that figure and we see that, though steady states of the branches AB and CD must be unstable, there can be no conclusion about the stability of solutions on the branch BC. There is in fact experimental evidence that these can sometimes be stable.

7.3. Methods based on the maximum principle and other comparison theorems

It should be clear from all that we have done that a great simplification obtains if $\mathscr{L} = 1$, and we will continue to describe the work that has been done under this assumption in spite of its lack of realism. Apart from showing some useful methods, it is an important base case from which we can move, and it is the situation in which we can go farthest toward answering the questions raised in Section 7.1. The key result in this case was the demonstration by Luss and Lee (1968) that a unique steady state is stable: this has been followed by some more extended results of Amundson and Varma (1972*a*).

We will again proceed on the basis of constant residual enthalpy. If we show that a unique steady state is stable and its domain of attraction is the whole state space then, the effect of residual enthalpy may be neglected. For the solution of the general equations analogous to eqns (7.22) and (7.24) has the form

$$z(\boldsymbol{\rho}, \tau) = 1+\beta+\sum_{n=1}^{\infty} a_n z_n(\boldsymbol{\rho}) \exp(-\lambda_n \tau),$$

where the coefficients a_n have to be chosen to satisfy the initial distribution $z_0(\boldsymbol{\rho}) = \beta u_0(\boldsymbol{\rho})+v_0(\boldsymbol{\rho})$. But whatever this initial distribution may be it is always possible to follow the state with the full equations until after time T we find that

$$|z(\boldsymbol{\rho}, \tau)-1-\beta| < \varepsilon, \qquad \boldsymbol{\rho} \in \Omega, \qquad \tau > T.$$

Now ε can be chosen arbitrarily small and the function $F(v, z)$ of eqn (7.23) will then differ arbitrarily little from $F(v)$ of eqn (7.26). If we can show that the unique steady state attracts from the initial condition $v(\boldsymbol{\rho}, T)$ and that the arbitrarily small deviation in F is not important, then the global stability will be established. If there are several steady states then, of course, the initial distribution of residual enthalpy will govern the steady state to which the transient solution tends.

The maximum principle for parabolic equations which is needed (see Il'lin, Kalashnikov, and Oleinik (1962)) can be adapted to our situation as follows: if $U(\boldsymbol{\rho}, \tau)$ is a continuous function satisfying

$$\frac{\partial U}{\partial \tau} = \nabla^2 U + c(\boldsymbol{\rho})U$$

in Ω and $c(\boldsymbol{\rho})$ is bounded above, then $U(\boldsymbol{\rho}, 0) \geqslant 0$ in Ω and $U(\boldsymbol{\rho}, \tau) \geqslant 0$ on $\partial\Omega$ imply that $U(\boldsymbol{\rho}, \tau) \geqslant 0$ in Ω for all $\tau > 0$.

Let $v_e(\boldsymbol{\rho})$ be the unique steady state solution of

$$\nabla^2 v_e + \beta\phi^2 F(v_e) = 0 \quad \text{in } \Omega \tag{7.45}$$

subject to

$$(v_e)_s = 1 \quad \text{on } \partial\Omega, \tag{7.46}$$

then

$$v^*(\boldsymbol{\rho}, \tau) = v(\boldsymbol{\rho}, \tau) - v_e(\boldsymbol{\rho}) \tag{7.47}$$

satisfies

$$\frac{\partial v^*}{\partial \tau} = \nabla^2 v^* + \beta\phi^2 \mathscr{F}(v_e, v^*)v^* \quad \text{in } \Omega \tag{7.48}$$

with

$$(v^*)_s = 0 \quad \text{on } \partial\Omega, \tag{7.49}$$

when

$$\mathscr{F} = \int_0^1 F'(v_e + \xi v^*)\,\mathrm{d}\xi. \tag{7.50}$$

Now $\mathscr{F}$, though unknown, is certainly bounded above so that the theorem applies to eqn (7.48), and if we can show that v^* is initially one-sided (i.e. either non-negative or non-positive) then it remains one-sided. Moreover, let

$$V^* = \frac{\partial v^*}{\partial \tau} = \frac{\partial v}{\partial \tau} \tag{7.51}$$

so that

$$\frac{\partial V^*}{\partial \tau} = \nabla^2 V^* + \beta\phi^2 \mathscr{F}'\{v(\boldsymbol{\rho}, \tau)\} V^* \quad \text{in } \Omega \tag{7.52}$$

and

$$V^* = 0 \quad \text{on } \partial\Omega. \tag{7.53}$$

Then the theorem applies also to this equation and the rate of change V^* enjoys the same one-sided behaviour.

Consider now the solution of the transient equations for which the initial distribution is $v_0(\boldsymbol{\rho}) \equiv 1$ in Ω and denote the corresponding deviation by v_1^*. Now v_1^* satisfies eqn (7.48) and

$$v_1^*(\boldsymbol{\rho}, 0) = 1 - v_e(\boldsymbol{\rho}) \leqslant 0$$

whereas

$$v_1^*(\boldsymbol{\rho}, \tau) = 0 \quad \text{if } \boldsymbol{\rho} \in \partial\Omega.$$

Thus by the maximum principle $v_1^*(\boldsymbol{\rho}, \tau)$ is never positive and v_1^* always lies below v_e. Also its rate of change

$$V_1^*(\boldsymbol{\rho}, 0) = \beta\phi^2 F(1)$$

is everywhere positive at $\tau = 0$ and zero on $\partial\Omega$. Hence $V_1^*(\boldsymbol{\rho}, \tau)$ is never negative and $v_1(\boldsymbol{\rho}, \tau)$ is monotone increasing in time at each point. It must converge to a steady state and since $v_e(\boldsymbol{\rho})$ is unique it must approach it from below. An entirely similar argument can be used for $v_{1+\beta}(\boldsymbol{\rho}, \tau)$ the solution of the transient equation whose initial value is constant and equal to $(1+\beta)$. It will therefore tend downwards toward the steady state. Figure 7.4, based

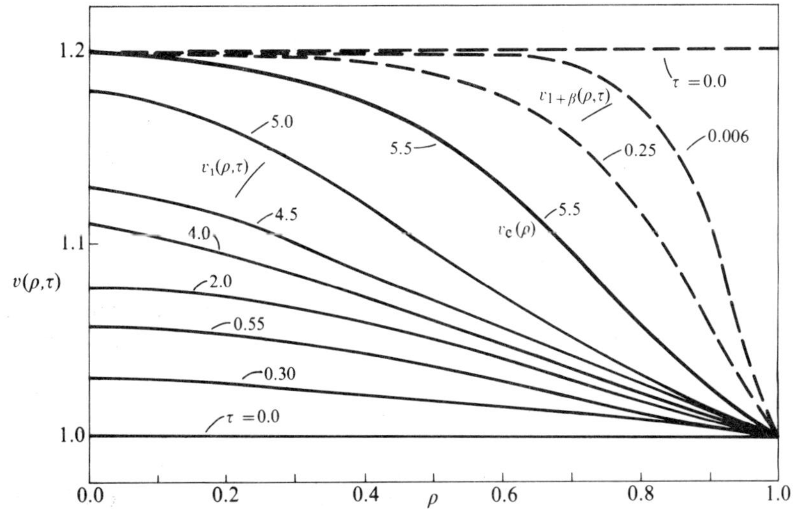

FIG. 7.4. Transient temperature profiles $v_1(\rho, \tau)$ and $v_{1+\beta}(\rho, \tau)$ for an exothermic first order reaction in a sphere; $\phi = 0{\cdot}86$, $\beta = 0{\cdot}2$, $\gamma = 30$. (After Luss and Lee (1968).)

on calculations of Luss and Lee, shows the way in which these two solutions approach $v_e(\boldsymbol{\rho})$ in a spherical pellet; it is noticeable that $v_{1+\beta}$ approaches v_e much more rapdily than v_1. Finally one can show in the same way that no solution with initial values $v_0(\boldsymbol{\rho})$ between 1 and $1+\beta$ ever crosses v_1

or $v_{1+\beta}$; that is

$$1 \leqslant v_0(\boldsymbol{\rho}) \leqslant 1+\beta$$

implies

$$v_1(\boldsymbol{\rho}, \tau) \leqslant v(\boldsymbol{\rho}, \tau) \leqslant v_{1+\beta}(\boldsymbol{\rho}, \tau). \tag{7.54}$$

Since v_1 and $v_{1+\beta}$ both converge to the steady state and they pinch any other transient between them, it follows that all solutions tend to the steady state and uniqueness implies global stability when $\mathscr{L} = 1$.

When there are three solutions we know from the arguments of the last paragraph that the middle one is unstable. The maximal and minimal solutions cannot intersect any other solution, so that when there are only three solutions none of them intersect one another except at $\rho = 1$. Let the minimal, unstable and maximal solutions be denoted by v_{e1}, v_{e2} and v_{e3} respectively. Then it can be shown by the same method as before that $v_1(\boldsymbol{\rho}, \tau)$ will converge from below on $v_{e1}(\boldsymbol{\rho})$ and $v_{1+\beta}(\boldsymbol{\rho}, \tau)$ will converge on $v_{e3}(\boldsymbol{\rho})$ from above. We now wish to show that any solution which starts with $v_0(\boldsymbol{\rho})$ everywhere above $v_{e2}(\boldsymbol{\rho})$ will tend to $v_{e3}(\boldsymbol{\rho})$, whereas one that starts below $v_{e2}(\boldsymbol{\rho})$ will tend to $v_{e1}(\boldsymbol{\rho})$. Let $w(\boldsymbol{\rho})$ be any bounded positive function vanishing on $\partial\Omega$ and $\varepsilon > 0$ then

$$v_{0+}(\boldsymbol{\rho}) = v_{e2}(\boldsymbol{\rho}) + \varepsilon w(\boldsymbol{\rho})$$

lies everywhere above v_{e2} and if ε is taken to be sufficiently small it will lie below v_{e3}. Thus $v_+(\boldsymbol{\rho}, \tau)$ the solution with this initial condition will always remain above v_{e2} and below v_{e3}. But if ε is small $v_+^*(\boldsymbol{\rho}, \tau) = v_+(\boldsymbol{\rho}, \tau) - v_{e2}(\boldsymbol{\rho})$ satisfies

$$\frac{\partial v_+^*}{\partial \tau} = \nabla^2 v_+^* + \beta\phi^2 F'(v_{e2}(\boldsymbol{\rho}))v^*$$

for small times. We know from the previous paragraph that this equation has a negative eigenvalue corresponding to instability. This will dominate and v_+^* will be approximately

$$v_+^* = \varepsilon w(\boldsymbol{\rho}) \exp(-\lambda_1 \tau), \quad \lambda_1 < 0$$

Hence V_+^* is positive and $v_+(\boldsymbol{\rho}, \tau)$, the solution with $v_+(\boldsymbol{\rho}, \tau) = v_{0+}(\boldsymbol{\rho})$, will tend upwards to $v_{e3}(\boldsymbol{\rho})$. A similar argument with an initial distribution $v_{0-}(\boldsymbol{\rho}) = v_{e2}(\boldsymbol{\rho}) - \varepsilon w(\boldsymbol{\rho})$ shows that $v_-(\boldsymbol{\rho}, \tau)$, the solution with this initial distribution, tends downwards to $v_{e1}(\boldsymbol{\rho})$. The region of attraction of $v_{e1}(\boldsymbol{\rho})$ includes all initial distributions satisfying

$$1 \leqslant v_0(\boldsymbol{\rho}) < v_{e2}(\boldsymbol{\rho}),$$

whilst that of $v_{e3}(\boldsymbol{\rho})$ includes all $v_0(\boldsymbol{\rho})$ satisfying

$$v_{e2}(\boldsymbol{\rho}) < v_0(\boldsymbol{\rho}) \leqslant 1+\beta.$$

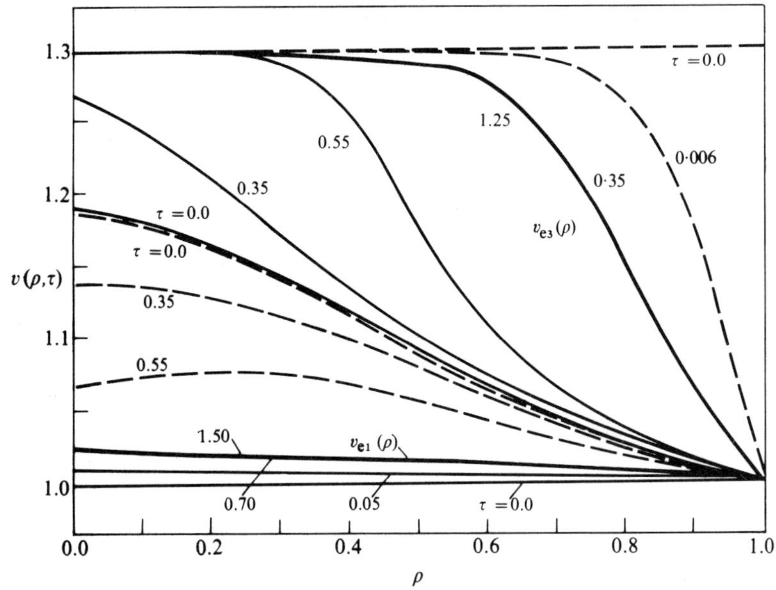

FIG. 7.5. Transient temperature profiles $v_1(\rho, \tau)$, $v_-(\rho, \tau)$, $v_+(\rho, \tau)$ and $v_{1+\beta}(\rho, \tau)$ for exothermic first order reaction in a sphere; $\phi = 0{\cdot}545$, $\beta = 0{\cdot}3$, $\gamma = 30$. (After Luss and Lee (1968).)

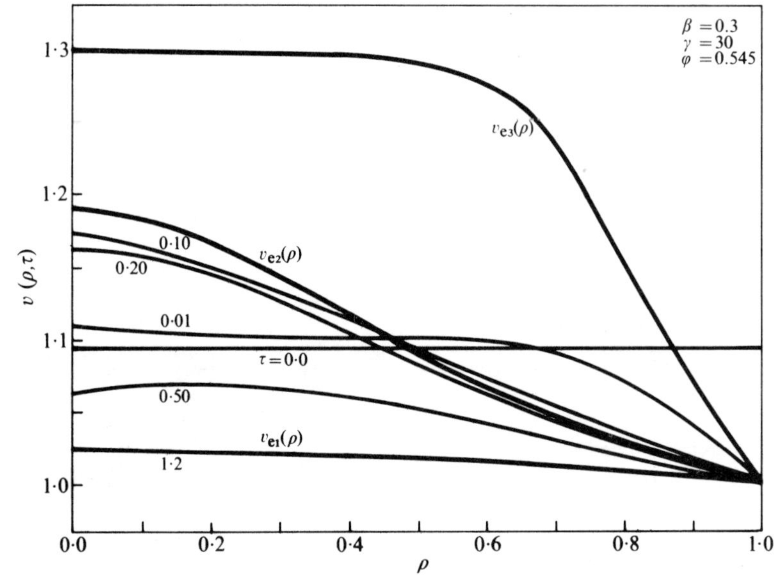

FIG. 7.6. Transient temperature profiles for when $v_0(\rho) = 1{\cdot}094$ under the same conditions as Fig. 7.5. (After Luss and Lee (1968).)

Figure 7.5 (based as are Figs. 7.4–7 on Luss and Lee (1968)) shows the transients $v_1(\rho, \tau)$, $v_-(\rho, \tau)$, $v_+(\rho, \tau)$, and $v_{1+\beta}(\rho, \tau)$ for an exothermic first-order reaction in a sphere; v_{0+} and v_{0-} are calculated with $w(\rho) = 1$, $\varepsilon = 0{\cdot}0005$.

Figure 7.6 is based on some numerical experiments by Luss and Lee to find where the separatrix lies in the class of constant functions. With $v_0(\rho) = 1{\cdot}094$ the solution ultimately tends to $v_{e1}(\rho)$ though $v(\rho, \tau)$ goes quite close to $v_{e2}(\rho)$ when τ is between 0·1 and 0·2. On the other hand the same parameter values with $v_0(\rho) = 1{\cdot}095$ lead the solution to $v_{e3}(\rho)$. Thus there is some constant value for $v_0(\rho)$ between 1·094 and 1·095 which lies in the region of attraction of $v_{e2}(\rho)$, the separatrix in $\mathscr{S}$ between the regions of attraction of the maximal solution. Of course even if this value could be found exactly the transient could never be computed, since the slightest round-off error would serve to tip the transient off the separatrix and lead to either v_{e1} or v_{e3}. Finally the effect of residual enthalpy is shown in Fig. 7.8 which is for exactly the same parameter values and initial distribution as Fig. 7.7. However because the residual enthalpy is not initially constant and equal to $(1+\beta)$, the solution goes to the minimal rather than the maximal steady state.

The same method can be used to show that the maximal and minimal solutions are stable when there are more than three solutions. For as before $v_{1+\beta}(\boldsymbol{\rho}, \tau)$ descends monotonically on $\hat{v}(\boldsymbol{\rho})$, the maximal solution, and if

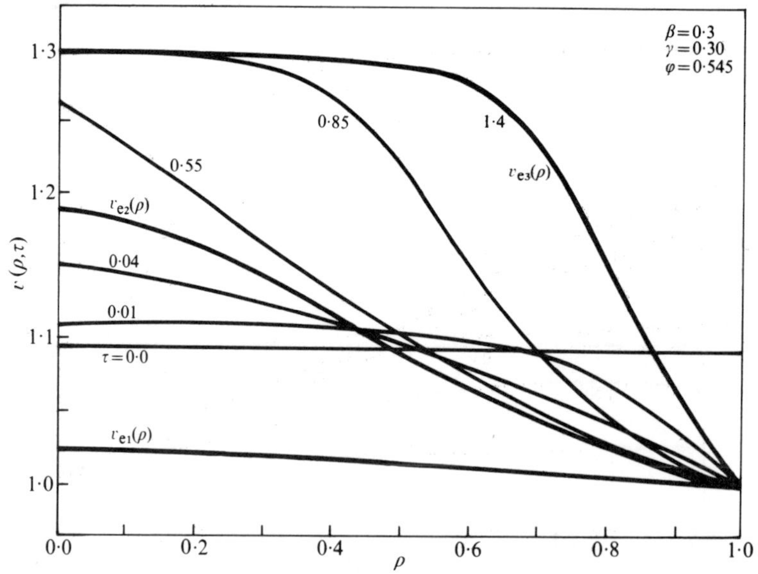

FIG. 7.7. Transient temperature profiles when $v_0(\rho) = 1{\cdot}095$ under the same conditions as Fig. 7.5. (After Luss and Lee (1968).)

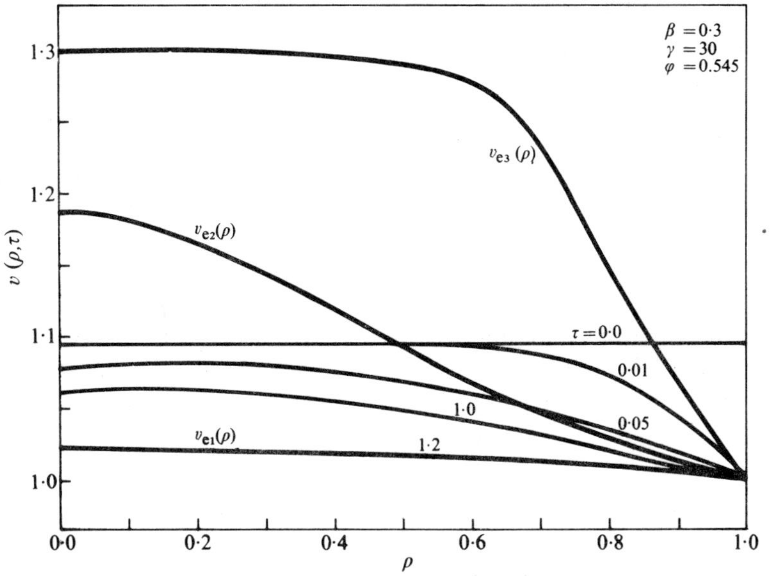

FIG. 7.8. Transient temperature profiles with the same initial conditions and parameter values as Fig. 7.7 but with residual enthalpy $z_0(\rho) = -0{\cdot}205$. (After Luss and Lee (1968).)

we can show that any solution starting close to $\hat{v}(\mathbf{\rho})$ but entirely below it tends upwards toward it then the 'pinching' argument will show that any initial conditions close to $\hat{v}(\mathbf{\rho})$ will lead to it. With $w(\mathbf{\rho})$ again defined as a positive function with zero boundary values we set

$$v_0(\mathbf{\rho}) = \hat{v}(\mathbf{\rho}) - \varepsilon w(\mathbf{\rho})$$

and calculate V_0^* the initial value of $\partial v/\partial \tau$. By the equation for v it is

$$\begin{aligned} V_0^* &= \nabla^2 v_0 + \beta\phi^2 F(v_0) \\ &= -\varepsilon\nabla^2 w - \beta\phi^2\{F(\hat{v}) - F(\hat{v} - \varepsilon w)\} \\ &= -\varepsilon\{\nabla^2 w + \beta\phi^2 F'(\hat{v})w\} \end{aligned}$$

if ε is sufficiently small. But

$$\nabla^2 w + \beta\phi^2 F'(\hat{v})w < -\lambda w, \quad \lambda > 0,$$

for if this were not so the maximal steady state would lie on a branch of the (η, ϕ)-curve for which the slope tends to $-\infty$ at a bifurcation point. Hence $V_0^* > \varepsilon\lambda w \geqslant 0$ and the solution $v(\mathbf{\rho}, \tau)$ must tend upwards from below. Similarly the minimal solution is stable, for it can be shown that, with $v_0(\mathbf{\rho}) > \check{v}(\mathbf{\rho})$, the solution will tend downwards from above. Also $v_1(\mathbf{\rho}, \tau)$ tends upwards from below. This concurs with the approach of Sattinger (1972) who has shown by Leray–Schauder theory that a stable solution remains stable throughout the branch and can only change its character at a bifurcation point.

A more precise account of the effect of residual enthalpy and of the nature of the approach to the steady state is given by Amundson and Varma (1972*a*) who used the so-called Westphal–Prodi–Szarski theorem. This can be stated rather generally for a nonlinear equation of parabolic type,

$$\frac{\partial U}{\partial \tau} = G\left(\boldsymbol{\rho}, \tau, U, \frac{\partial U}{\partial \rho_i}, \frac{\partial^2 U}{\partial \rho_i \partial \rho_j}\right). \tag{7.55}$$

It will be convenient to abbreviate the derivatives by the use of suffixes; thus $U_i = \partial U/\partial \rho_i$ and $U_{ij} = \partial^2 U/\partial \rho_i \partial \rho_j$. Then G is required to be continuous in all its arguments with continuous derivatives $\partial G/\partial U_{ij}$ such that

$$\sum_i \sum_j (\partial G/\partial U_{ij})\xi_i \xi_j$$

is a positive definite quadratic form in the real variables ξ. It is convenient to define certain regions of $(\boldsymbol{\rho}, \tau)$-space more precisely. If Ω is the interior of the catalyst particle regarded as an open set, $\bar{\Omega} = \Omega \cup \partial\Omega$ is its closure. The set of points $(\boldsymbol{\rho}, \tau)$ in $R^3 \times R$ such that $\boldsymbol{\rho} \in \bar{\Omega}$, $0 \leqslant \tau \leqslant T$ will be denoted by $\bar{\Omega}_{0,T}$ and $\Omega_{0,T}$ will denote the interior points of this hypercylinder (i.e. $(\boldsymbol{\rho},\tau)$) such that $\boldsymbol{\rho} \in \Omega$, $(0 < \tau < T)$. It is natural to put $\bar{\Omega}_{0,T} = \Omega_{0,T} \cup \partial\Omega_{0,T}$ where $\partial\Omega_{0,T}$ consists of all the points $(\boldsymbol{\rho}, \tau)$ with $\boldsymbol{\rho} \in \partial\Omega$ and $0 \leqslant \tau \leqslant T$. If we want to call attention to the section of this hypercylinder by a plane of constant τ, i.e. the points $(\boldsymbol{\rho}, \tau)$ with τ fixed, then we can refer to Ω_τ or $\bar{\Omega}_\tau$. These definitions are illustrated in Fig. 7.9. Consider two functions $U(\boldsymbol{\rho}, \tau)$ and $W(\boldsymbol{\rho}, \tau)$ such that

$$\begin{aligned} U_\tau &\leqslant G(\boldsymbol{\rho}, \tau, U, U_i, U_{ij}), \\ W_\tau &> G(\boldsymbol{\rho}, \tau, W, W_i, W_{ij}) \end{aligned} \tag{7.56}$$

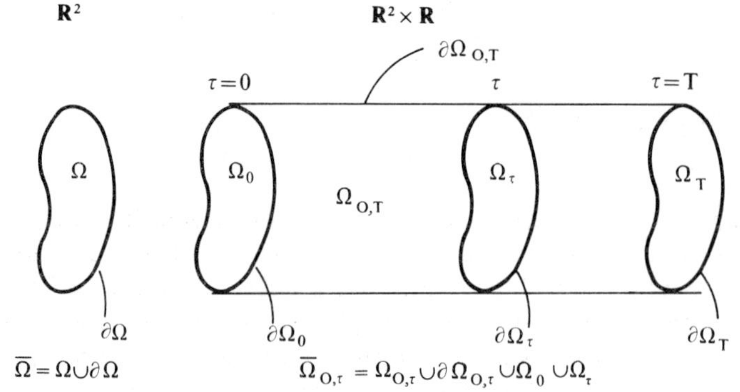

FIG. 7.9. The regions involved in the Westphal–Prodi–Szarski theorem.

in $\Omega_{0,T} \cup \Omega_T$. If

$$W(\boldsymbol{\rho}, \tau) > U(\boldsymbol{\rho}, \tau) \quad \text{on } \bar{\Omega}_0 \cup \partial\Omega_{0,T} \tag{7.57}$$

then

$$W(\boldsymbol{\rho}, \tau) > U(\boldsymbol{\rho}, \tau) \quad \text{in } \bar{\Omega}_{0,T}. \tag{7.58}$$

Notice that the inequalities (7.56) imply

$$W_\tau - G(\boldsymbol{\rho}, \tau, W, W_i, W_{ij}) > U_\tau - G(\boldsymbol{\rho}, \tau, U, U_i, U_{ij}). \tag{7.59}$$

This leads to a stability criterion of the following form. If $V(\boldsymbol{\rho}, \tau)$ satisfies

$$\frac{\partial V}{\partial \tau} = \nabla^2 V + h(\boldsymbol{\rho}, \tau, V)V \quad \text{in } \Omega_{0,T} \tag{7.60}$$

with

$$V = 0 \qquad \text{on } \partial\Omega_{0,T} \tag{7.61}$$

and

$$V = V_0(\boldsymbol{\rho}) \quad \text{on } \bar{\Omega}_0 \tag{7.62}$$

then the null solution, $V \equiv 0$, of eqn (7.60) will be exponentially asymptotically stable if $V_0(\boldsymbol{\rho})$ is bounded on $\bar{\Omega}_0$ and

$$\sup h(\boldsymbol{\rho}, \tau, V) < \lambda_1^2, \tag{7.63}$$

the least eigenvalue of the equation

$$\nabla^2 w + \lambda^2 w = 0 \quad \text{in } \Omega, \qquad w = 0 \quad \text{on } \partial\Omega. \tag{7.64}$$

The supremum in eqn (7.63) is to be taken for all $(\boldsymbol{\rho}, \tau) \in \Omega_{0,T} \cup \Omega_T$ and for all admissible values of U. The proof uses a comparison function based on the first eigenfunction of the Laplacian for a region slightly larger than Ω. Let $\Omega_\alpha \supset \Omega$ denote a region enclosing Ω entirely with no common boundary points (i.e. $\partial\Omega \cap \partial\Omega_\alpha = \varnothing$), then the first eigenvalue of the Laplacian in Ω_α is slightly less than that of Ω and we write

$$\nabla^2 w_1 + (\lambda_1^2 - \alpha) w_1 \quad \text{in } \Omega_\alpha, \qquad w_1 = 0 \quad \text{on } \partial\Omega_\alpha.$$

Moreover w_1 is positive in Ω_α and so bounded below away from zero in Ω and on $\partial\Omega$. Let

$$W(\boldsymbol{\rho}, \tau) = A w_1(\boldsymbol{\rho})\, \mathrm{e}^{-\alpha\tau} \tag{7.65}$$

and choose A so large that $Aw_1(\boldsymbol{\rho}) > V_0(\boldsymbol{\rho})$ for all $\boldsymbol{\rho} \in \bar{\Omega}$. Then $W(\boldsymbol{\rho}, \tau)$ has the properties

(i) $W > V \quad \text{on } \bar{\Omega}_0 \cup \partial\Omega_{0,T}$
(ii) $W > 0 \quad \text{in } \bar{\Omega}_{0,T}$
(iii) $W(\boldsymbol{\rho}, \tau) \to 0 \quad \text{as } \tau \to \infty \quad \text{in } \bar{\Omega}_{0,\tau}$.

Also

$$W_\tau - \nabla^2 W - h(\boldsymbol{\rho}, \tau, V)W = W\{\lambda_1^2 - 2\alpha - h(\boldsymbol{\rho}, \tau, V)\}$$
$$\geqslant W\{\lambda_1^2 - 2\alpha - \sup h\} > 0$$

if α is chosen sufficiently small. Hence if we identify V with U and $\nabla^2 V + h(\boldsymbol{\rho}, \tau, V)V$ with the G of the Westphal–Prodi–Szarski theorem we see that

$$V(\boldsymbol{\rho}, \tau) \leqslant W(\boldsymbol{\rho}, \tau) \quad \text{in } \bar{\Omega}_{0,\tau}$$

and hence tends to zero at least as rapidly as $\exp(-\alpha\tau)$. A comparison function $-W$ can also be used and by a similar argument

$$-W(\boldsymbol{\rho}, \tau) \leqslant V(\boldsymbol{\rho}, \tau) \quad \text{in } \bar{\Omega}_{0,\tau},$$

or, more compactly

$$|V(\boldsymbol{\rho}, \tau)| \leqslant W(\boldsymbol{\rho}, \tau) \quad \text{in } \bar{\Omega}_{0,\tau} \tag{7.66}$$

Moreover A can be chosen large enough to accommodate any bounded initial conditions so that the null state $V \equiv 0$ is globally, exponentially and asymptotically stable.

We now apply this stability theorem to the system of equations for v^*, the deviation of the temperature from steady state, and z^*, the deviation of the residual enthalpy from the constant $1+\beta$. We have

$$\frac{\partial v^*}{\partial \tau} = \nabla^2 v^* + \beta\phi^2\{F(v_e + v^*, 1+\beta+z^*) - F(v_e)\}, \tag{7.67}$$

$$\frac{\partial z^*}{\partial \tau} = \nabla^2 z^* \tag{7.68}$$

in Ω with $v^* - z^* = 0$ on $\partial\Omega$. $F(v, z)$ and $F(v)$ are defined for the pth-order reaction in eqns (7.23) and (7.26). The equation for z^* can be solved in terms of the eigenfunctions of eqn (7.64),

$$z^*(\boldsymbol{\rho}, \tau) = \sum a_n w_n(\boldsymbol{\rho}) \exp(-\lambda_n^2 \tau), \tag{7.69}$$

where the a_n are chosen to match the initial distribution of residual enthalpy. When the residual enthalpy is initially constant and equal to $(1+\beta)$ all the a_n are zero and we can write the nonlinear term in eqn (7.67) as

$$h(\boldsymbol{\rho}, \tau, v^*) = \beta\phi^2 F'(v_e + \theta v^*),$$

where $0 \leqslant \theta \leqslant 1$. Then the equation for v^* is of the form of eqn (7.60) and stability is assured if

$$\beta\phi^2 \max[F'(v_e)] < \lambda_1^2. \tag{7.70}$$

If the solution v_e is inserted and the maximum calculated this condition ensures its stability. A sufficient condition for any solution to be stable is

$$\beta\phi^2 F'(v_i) < \lambda_1^2, \tag{7.71}$$

where v_i is the value of v at the point of inflection.

If the residual enthalpy is not initially constant we have to work a little harder. The nonlinear term in eqn (7.67) must be written

$$\{\beta\phi^2 F'(v_e + \theta v^*)\}v^* + \{\beta\phi^2 F_z(v_e, 1+\beta+\theta' z^*)\}z^*,$$

where F_z denotes the derivative of $F(z, v)$ with respect to z and $0 \leqslant \theta, \theta' \leqslant 1$. In this case the constant A in the comparison function W must be chosen so large that both

$$Aw_1(\boldsymbol{\rho}) > V_0(\boldsymbol{\rho})$$

and

$$A\{\lambda_1^2 - 2\alpha - \beta\phi^2 F'(v_e + \theta v^*)\}w_1(\boldsymbol{\rho}) > \{\beta\phi^2 \sup|F_z|\}|z_0^*(\boldsymbol{\rho})|.$$

It will then be again found that $W_\tau - G > 0$ and the comparison goes through. Since this is always possible provided the initial conditions are bounded and eqn (7.70) is satisfied, it is clear that this equation gives a sufficient condition for the stability of the steady state. We notice that the steady state for an endothermic reaction with a monotonic $F(v)$, which we know is unique, is also stable since the left hand side of eqn (7.70) is then negative and so the inequality always satisfied.

7.4. Liapounov's direct method

Wei (1965) was the first to suggest that the Liapounov method might be applied to the stability problem of the catalyst particle. He obtained the sufficient condition $\beta\gamma < 1$ and a special case of the condition obtained in the last paragraph. Since then there have been several papers discussing the problem as the references at the end of the chapter will show, we shall follow the route laid out by Lapidus, Padmanabhan, and Yang (1971) in the case of the Dirichlet problem for a single reaction.

We may start by adapting eqns (7.6) and (7.7) to the present case, taking $\Delta_1 = 1, \alpha_1 = -1$ and dropping the suffix on u. Thus the deviations from the steady state $u_e(\boldsymbol{\rho}), v_e(\boldsymbol{\rho})$ satisfy

$$\frac{\partial u^*}{\partial \tau} = \nabla^2 u^* - \phi^2 R^*(u^*, v^*, \boldsymbol{\rho}), \tag{7.72}$$

$$\mathscr{L}\frac{\partial v^*}{\partial \tau} = \nabla^2 v^* + \beta\phi^2 R^*(u^*, v^*, \boldsymbol{\rho}), \tag{7.73}$$

in Ω with

$$u^* = v^* = 0 \tag{7.74}$$

on $\partial\Omega$. Though the method is applicable when $\mathscr{L} \neq 1$ it will be well to illustrate it first by the simpler case $\mathscr{L} = 1$. Here we can take eqns (7.67) and (7.68) and work with v^* and z^* the deviation of the residual enthalpy. Let us write then

$$\frac{\partial v^*}{\partial \tau} = \nabla^2 v^* + \beta\phi^2 \mathscr{F}(v^*, z^*, \boldsymbol{\rho}), \tag{7.75}$$

$$\frac{\partial z^*}{\partial \tau} = \nabla^2 z^* \tag{7.76}$$

and consider the functional of the solution defined by

$$V(\tau) = \tfrac{1}{2} \iiint_\Omega \{v^{*2} + \chi^2 z^{*2}\}\, \mathrm{d}\Upsilon \tag{7.77}$$

where χ^2 is a positive constant to be chosen later. $V(\tau)$ is clearly a positive function which vanishes only at the steady state under scrutiny. If we can show that its derivative is always negative except when $v^* = z^* = 0$, then it will decrease to zero as $\tau \to \infty$ and the steady state in question will be stable. Now

$$V'(\tau) = \iiint_\Omega \left(v^* \frac{\partial v^*}{\partial \tau} + \chi^2 z^* \frac{\partial z^*}{\partial \tau} \right) \mathrm{d}\Upsilon \tag{7.78}$$

and by substitution from eqns (7.75) and (7.76) this can be written

$$V'(\tau) = \iiint_\Omega (v^* \nabla^2 v^* + \chi^2 z^* \nabla^2 z^*)\, \mathrm{d}\Upsilon + \\ + \beta\phi^2 \iiint_\Omega \{v^* \mathscr{F}(v^*, z^*, \boldsymbol{\rho})\}\, \mathrm{d}\Upsilon. \tag{7.79}$$

Now, if u^* is any square integrable function vanishing on $\partial\Omega$, we have the inequality

$$\iiint_\Omega (u^* \nabla^2 u^*)\, \mathrm{d}\Upsilon \leqslant -\lambda_1^2 \iiint_\Omega u^{*2}\, \mathrm{d}\Upsilon, \tag{7.80}$$

where λ_1^2 is the least eigenvalue of eqn (7.64). Hence

$$V'(\tau) \leqslant -\left[\iiint_\Omega \{\lambda_1^2 v^{*2} + \lambda_1^2 \chi^2 z^{*2} - \beta\phi^2 \mathscr{F}(v^*, z^*, \boldsymbol{\rho}) v^*\}\, \mathrm{d}\Upsilon \right] \tag{7.81}$$

and if we can find conditions under which the integrand is positive we shall have established stability.

We note in passing that the endothermic reaction with positive $\mathscr{F}$ always gives a stable steady state. Also the integrand can always be made positive by making ϕ sufficiently small. If we are considering only local stability $\mathscr{F}$

can be expanded about the origin and the higher powers of v^* and z^* can be neglected. Thus

$$\mathscr{F}(v^*, z^*, \boldsymbol{\rho}) \doteqdot \mathscr{F}_v(\boldsymbol{\rho})v^* + \mathscr{F}_z(\boldsymbol{\rho})z^*,$$

where

$$\mathscr{F}_v(\boldsymbol{\rho}) = R_v - \frac{1}{\beta}R_u, \qquad \mathscr{F}_z(\boldsymbol{\rho}) = \frac{1}{\beta}R_u$$

and the derivatives of R are evaluated on the steady state $u_e(\boldsymbol{\rho})$, $v_e(\boldsymbol{\rho})$ which is being studied. Then the inequality for $V'(\tau)$ can be written

$$V'(\tau) \leqslant -\left[\iiint_{\Omega} \{(\lambda_1^2 + \phi^2 R_u - \beta\phi^2 R_v)v^{*2} + R_u v^* z^* + \chi^2\lambda_1^2 z^{*2}\}\,\mathrm{d}\Upsilon\right] \tag{7.82}$$

We can ensure that this quadratic is positive provided only that the coefficient of v^{*2} is positive, for

$$\chi^2 > [\sup(\lambda_1^2 + \phi^2 R_u - \beta\phi^2 R_v)]/4\lambda_1^2[\inf R_u] \tag{7.83}$$

will ensure that the discriminant is negative. This condition on the coefficient of v^{*2} is just the same as eqn (7.70) for

$$\beta\phi^2 F'(v_e) = \beta\phi^2 R_v - \phi^2 R_u < \lambda_1^2, \tag{7.84}$$

again evaluated for the steady state under consideration. The sufficient condition (7.71) can also be derived from eqn. (7.84). For the pth-order irreversible reaction we may summarize the stability information inherent in eqn (7.84) as follows: any solution will be stable provided that

$$\phi^2 < \begin{cases} \infty, & \text{if } \beta\gamma \leqslant p \\ \lambda_1^2/(\beta\gamma - p), & \text{if } p \leqslant \beta\gamma \leqslant p + \beta + \{p + 2p\beta + \beta^2\}^{\frac{1}{2}} \\ \lambda_1^2/\beta F'(v_i), & \text{if } \beta\gamma \geqslant p + \beta + \{p + 2p\beta + \beta^2\}^{\frac{1}{2}} \end{cases} \tag{7.85}$$

It is only for the first-order reaction that a closed expression can be found for $F'(v_i)$ the slope at the point of inflection. We notice that the first sufficient condition, which ensures stability for all ϕ, is more restrictive then the condition for uniqueness. For example, with the first-order reaction $\beta\gamma \leqslant 1$ is more restrictive than $\beta\gamma \leqslant 4(1+\beta)$. This is a common phenomenon in the application of Liapounov functionals and arises from two causes. In the first place we have had to discard a lot of information in deriving these conditions. In the second, the choice of Liapounov functional may not have been the most appropriate and another choice might have led to a more liberal condition. We have not used any information about the particular solution being studied. Lapidus, Padmanabhan, and Yang (1971) show that certain general remarks can be made and that for symmetrical shapes rather simpler conditions can be obtained to ensure the positive definiteness of the integrand in the formula for $-V'(\tau)$.

When the Lewis number is not equal to one we let

$$z^* = \beta u^* + \mathscr{L} v^* \tag{7.86}$$

play the role of the residual enthalpy deviation. Then eqn (7.75), which now has a factor of $\mathscr{L}$ multiplying v^*_τ, must be coupled with

$$\frac{\partial z^*}{\partial \tau} = \nabla^2 z^* + (1-\mathscr{L})\nabla^2 v^* \tag{7.87}$$

and the form of $\mathscr{F}$ is changed slightly. If we linearize it for small disturbances v^*, z^*, the equation for v^* may be written

$$\mathscr{L}\frac{\partial v^*}{\partial \tau} = \nabla^2 v^* + g(\boldsymbol{\rho})v^* + h(\boldsymbol{\rho})z^*, \tag{7.88}$$

where

$$g = \phi^2(\beta R_v - \mathscr{L} R_u), \qquad h = \phi^2 R_u, \tag{7.89}$$

and the derivatives are evaluated along the steady state $u_e(\boldsymbol{\rho})$, $v_e(\boldsymbol{\rho})$. Now let

$$V(\tau) = \tfrac{1}{2}\iiint\limits_{\Omega} \{v^{*2} + \chi^2 z^{*2}\}\, \mathrm{d}\Upsilon,$$

so that

$$V'(\tau) = \iiint\limits_{\Omega} \{v^*\nabla^2 v^* + \chi^2 z^*\nabla^2 z^* + \chi^2(1-\mathscr{L})z^*\nabla^2 v^* + g v^{*2} + h z^* v^*\}\, \mathrm{d}\Upsilon.$$

Since v^* and z^* both vanish on $\partial\Omega$, Green's formula shows that

$$\iiint\limits_{\Omega} z^*\nabla^2 v^*\, \mathrm{d}\Upsilon = \iiint\limits_{\Omega} v^*\nabla^2 z^*\, \mathrm{d}\Upsilon$$

and $V'(\tau)$ may be written in a symmetrical form. Let $\mathbf{w}^*$ denote the column vector whose components are v^* and z^*. Then

$$V'(\tau) = \iiint\limits_{\Omega} (\mathbf{w}^*)^T(\mathbf{A}\nabla^2 + \mathbf{B})(\mathbf{w}^*)\, \mathrm{d}\Upsilon \tag{7.90}$$

where

$$\mathbf{A} = \begin{bmatrix} 1 & \tfrac{1}{2}\chi^2(1-\mathscr{L}) \\ \tfrac{1}{2}\chi^2(1-\mathscr{L}) & \chi^2 \end{bmatrix}, \qquad \mathbf{B} = \begin{bmatrix} g & \tfrac{1}{2}h \\ \tfrac{1}{2}h & 0 \end{bmatrix}, \tag{7.91}$$

and $V'(\tau)$ will be negative if the operator is negative definite.

Consider v^* and z^* to be expanded in terms of the eigenfunctions of eqn (7.64), i.e. $v^* = \sum a_n w_n(\boldsymbol{\rho})$ and $z^* = \sum b_n w_n(\boldsymbol{\rho})$, then

$$-\iiint_\Omega (\mathbf{w}^*)^T \mathbf{A}\nabla^2 \mathbf{w}^* \, \mathrm{d}\Upsilon = \sum \lambda_n^2 a_n^2 + \chi^2(1-\mathscr{L}) \sum \lambda_n^2 a_n b_n + \chi^2 \sum \lambda_n^2 b_n^2.$$

Let $A^2 = \sum \lambda_n^2 a_n^2$ and $B^2 = \sum \lambda_n^2 b_n^2$ then by Cauchy's inequality $(\sum \lambda_n^2 a_n b_n)^2 \leqslant A^2 B^2$ and

$$A^2 - \chi^2|1-\mathscr{L}|AB + \chi^2 B^2 \leqslant -\iiint_\Omega (\mathbf{w}^*)^T \mathbf{A}\nabla^2 \mathbf{w}^* \, \mathrm{d}\Upsilon$$

But

$$\begin{aligned} A^2 - \chi^2|1-\mathscr{L}|AB + \chi^2 B^2 &= (A - \tfrac{1}{2}\chi^2|1-\mathscr{L}|B)^2 + \chi^2\{1 - \tfrac{1}{4}\chi^2(1-\mathscr{L})^2\}B^2 \\ &= A^2\{1 - \tfrac{1}{4}\chi^2(1-\mathscr{L})^2\} + \chi^2(B - \tfrac{1}{2}|1-\mathscr{L}|A)^2 \\ &\geqslant \{1 - \tfrac{1}{4}\chi^2(1-\mathscr{L})^2\}\{(1-\omega)A^2 + \omega\chi^2 B^2\} \end{aligned}$$

for any ω, $0 \leqslant \omega \leqslant 1$. Moreover,

$$A^2 \geqslant \lambda_1^2 \sum a_n^2 = \lambda_1^2 \iiint (v^*)^2 \, \mathrm{d}\Upsilon$$

and

$$B^2 \geqslant \lambda_1^2 \sum b_n^2 = \lambda_1^2 \iiint (z^*)^2 \, \mathrm{d}\Upsilon$$

so that

$$-\iiint_\Omega (\mathbf{w}^*)^T \mathbf{A}\nabla^2 \mathbf{w}^* \, \mathrm{d}\Upsilon \geqslant \lambda_1^2\{1 - \tfrac{1}{4}\chi^2(1-\mathscr{L})^2\} \times$$
$$\times \iiint_\Omega \{(1-\omega)(v^*)^2 + \omega\chi^2(z^*)^2\} \, \mathrm{d}\Upsilon. \tag{7.92}$$

This excursus is necessary because there is no way of predicting the sign of $\sum \lambda_n^2 a_n b_n$ and so of getting a bound on AB in terms of the norms of v^* and z^*. Thus

$$-V'(\tau) \geqslant \iiint_\Omega [(\lambda_1^2\{1 - \tfrac{1}{4}\chi^2(1-\mathscr{L})^2\}(1-\omega) - g)(v^*)^2 + hv^*z^* +$$
$$+ \lambda_1^2\{1 - \tfrac{1}{4}\chi^2(1-\mathscr{L})^2\}\omega\chi^2(z^*)^2] \, \mathrm{d}\Upsilon,$$

and the integrand is positive definite if

$$\lambda_1^2 \mathscr{G} = \sup_{\boldsymbol{\rho}\in\Omega} g(\boldsymbol{\rho}) < \lambda_1^2\{1 - \tfrac{1}{4}\chi^2(1-\mathscr{L})^2\}(1-\omega) \tag{7.93}$$

and

$$\mathscr{H}^2 = \sup_{\boldsymbol{\rho}\in\Omega}[h(\boldsymbol{\rho})]^2 \leqslant 4\lambda_1^4\{1-\tfrac{1}{4}\chi^2(1-\mathscr{L})^2\}\omega\chi^2[\{1-\tfrac{1}{4}\chi^2(1-\mathscr{L})^2\}(1-\omega)-\mathscr{G}]. \tag{7.94}$$

The constants χ^2 and ω still remain to be chosen as judiciously as possible. Now eqn (7.93) cannot be satisfied for any permitted choice of χ^2 and ω unless

$$\mathscr{G} = [\sup g(\boldsymbol{\rho})]/\lambda_1^2 < 1. \tag{7.95}$$

If this condition is satisfied then χ and ω should be chosen to make the right-hand side of eqn (7.94) as large as possible. For this let

$$\omega = \frac{1-\mathscr{G}}{2+\mathscr{G}}, \qquad \chi^2 = \frac{4}{3}\frac{1-\mathscr{G}}{(1-\mathscr{L})^2} \tag{7.96}$$

and then eqn (7.94) becomes

$$\mathscr{H}^2 \leqslant \frac{16}{27}\frac{\lambda_1^4}{(1-\mathscr{L})^2}(1-\mathscr{G})^3. \tag{7.97}$$

Equations (7.95) and (7.97) are sufficient conditions for stability, the first being the same as eqn (7.70) or (7.84). If $\mathscr{L}$ is sufficiently close to 1 the first condition suffices of itself.

The conclusions that can be drawn from a Liapounov function are somewhat limited, since failure to find a positive definite V with a negative definite derivative may not be due to the instability of the system.

7.5. The effect of Lewis number on the stability of the steady state

In the preceding sections linearization has played an incidental, but increasingly important role. For the maximum principle it was not needed, whereas it was found to be a convenient way of developing conditions of monotonicity of the Liapounov functional. We now turn to linearization as a technique for discussing the stability of a given steady state. This may be most clearly illustrated by continuing with the example of an irreversible exothermic reaction: the more general case will be discussed in Section 7.6.

To recapitulate, we have a steady state solution $u_e(\boldsymbol{\rho})$, $v_e(\boldsymbol{\rho})$ satisfying

$$0 = \nabla^2 u_e - \phi^2 R(u_e, v_e), \tag{7.98}$$

$$0 = \nabla^2 v_e + \beta\phi^2 R(u_e, v_e) \tag{7.99}$$

in Ω, with

$$u_e = v_e = 1 \tag{7.100}$$

on $\partial\Omega$, while the deviations from steady state

$$u^*(\boldsymbol{\rho}, \tau) = u(\boldsymbol{\rho}, \tau) - u_e(\boldsymbol{\rho}), \qquad v^*(\boldsymbol{\rho}, \tau) = v(\boldsymbol{\rho}, \tau) - v_e(\boldsymbol{\rho}) \tag{7.101}$$

satisfy

$$\frac{\partial u^*}{\partial \tau} = \nabla^2 u^* - \phi^2 R^*(u^*, v^*, \boldsymbol{\rho}), \tag{7.102}$$

$$\mathscr{L}\frac{\partial v^*}{\partial \tau} = \nabla^2 v^* + \beta\phi^2 R^*(u^*, v^*, \boldsymbol{\rho}) \tag{7.103}$$

in Ω, with

$$u^* = v^* = 0 \tag{7.104}$$

on $\partial\Omega$. The method of linearization asserts that the local asymptotic stability of the steady state can be decided as follows. Let

$$u^*(\boldsymbol{\rho}, \tau) = U(\boldsymbol{\rho})\, \mathrm{e}^{-\mu\tau}, \qquad v^*(\boldsymbol{\rho}, \tau) = V(\boldsymbol{\rho})\, \mathrm{e}^{-\mu\tau}, \tag{7.105}$$

where μ is a complex number and U and V satisfy the linearized equations

$$\nabla^2 U + (\mu - \phi^2 R_u)U - (\phi^2 R_v)V = 0, \tag{7.106}$$

$$\nabla^2 V + (\beta\phi^2 R_u)U + (\mathscr{L}\mu + \beta\phi^2 R_v)V = 0 \tag{7.107}$$

in Ω, with

$$U = V = 0 \tag{7.108}$$

on $\partial\Omega$. In this equation the coefficients are functions of position $\boldsymbol{\rho}$ since the partial derivatives are evaluated along the steady state solutions. If the eigenvalues of the linear system (7.106–8) all have positive real parts then sufficiently small perturbations satisfying eqns (7.102–4) will die away and the steady state defined by eqns (7.98–100) will be locally asymptotically stable.

We are now faced with the eigenvalue problem for a pair of linear differential equations with variable coefficients and there are several methods of attack. The one which concerns us here is a direct numerical approach with suitable safeguards for accuracy. Such a method was used by Luss and Lee (1970) whose results are so important that they are worth giving in full. Briefly, the numerical method is to substitute for U and V the finite series approximations

$$U(\boldsymbol{\rho}) = \sum_{n=1}^{N} a_n w_n(\boldsymbol{\rho}), \qquad V(\boldsymbol{\rho}) = \sum_{n=1}^{N} b_n w_n(\boldsymbol{\rho}), \tag{7.109}$$

where $w_1(\boldsymbol{\rho}), \ldots, w_N(\boldsymbol{\rho})$ are the first N eigenfunctions of eqn (7.64). Then multiplying each equation by $w_m(\boldsymbol{\rho})$, and integration over Ω gives $2N$

homogeneous linear equations for the a_n and b_n:

$$\sum_{n=1}^{N}\left\{(\mu-\lambda_m^2)\delta_{mn}-\phi^2 A_{mn}\right\}a_n-\sum_{n=1}^{N}\phi^2 B_{mn}b_n=0,$$
$$\sum_{n=1}^{N}\beta\phi^2 A_{mn}a_n+\sum_{n=1}^{N}\left\{(\mathscr{L}\mu-\lambda_n^2)\delta_{mn}+\beta\phi_n^2 B_{mn}\right\}b_n=0, \tag{7.110}$$

where

$$A_{mn}=\iiint_{\Omega} R_n w_m w_n \,\mathrm{d}\Upsilon, \qquad B_{mn}=\iiint_{\Omega} R_v w_m w_n \,\mathrm{d}\Upsilon. \tag{7.111}$$

Setting the determinant of these equations equal to zero gives an algebraic equation for μ. The numerical safeguards usually applied are changes in mesh size for the integrations and in N, the number of terms in the approximation. These are increased until no significant change is observed in the required number of eigenvalues. Thus at the expense of some computer time, as much accuracy as is wanted can be obtained. Luss and Lee report the need for extreme accuracy with small values of $\mathscr{L}$.

The following results of Luss and Lee (1970) apply to the irreversible first-order reaction in a sphere for two cases of exothermic kinetics. The first case $\beta = 0{\cdot}15$, $\gamma = 30$ satisfies the condition $\beta\gamma < 4(1+\beta)$, so that we can be sure of a unique solution for all values of ϕ. The second case with $\beta = 0{\cdot}3$, $\gamma = 30$ will give multiple solutions for $0{\cdot}41 < \phi < 0{\cdot}66$, as the graph of $v_e(0)$ versus ϕ in Fig. 7.10 shows.

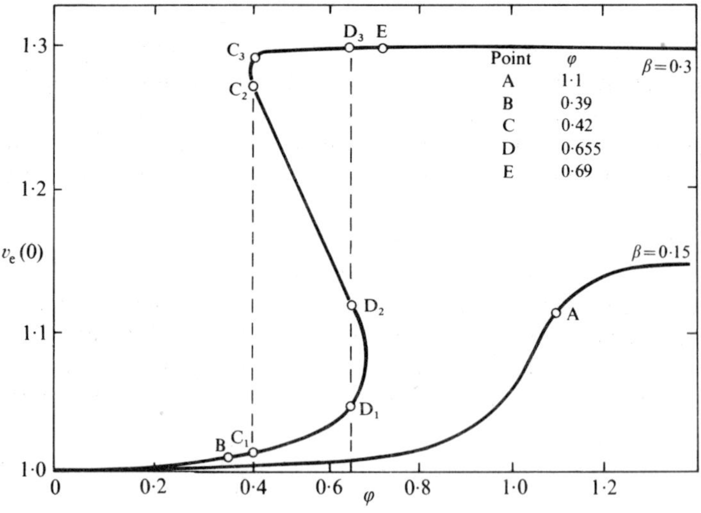

FIG. 7.10. Maximum temperature as a function Thiele modulus for $\gamma = 30$ and $\beta = 0{\cdot}3$, $0{\cdot}15$. (After Luss and Lee (1970).)

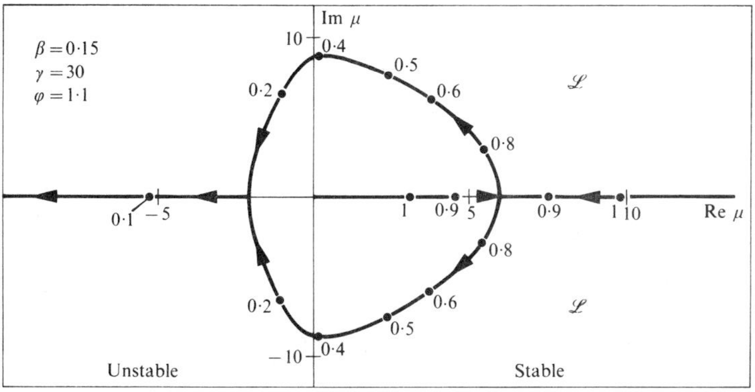

FIG. 7.11. Movement of the first two eigenvalues for the steady state *A* of Fig. 7.10. The numbers adjacent to dots are the values of $\mathscr{L}$.

In the first case a value of $\phi = 1{\cdot}1$ was chosen corresponding to the point A on the lower curve of Fig. 7.10. Luss and Lee determined the first 14 eigenvalues for $\mathscr{L}$ varying from 2·0 to 0·1. For very large $\mathscr{L}$ the least eigenvalue is positive and close to zero and as $\mathscr{L}$ decreases toward 1 it moves along the real axis. Since the least eigenvalue actually increases as $\mathscr{L}$ decreases the system does in a sense become more stable at first. However it has generally been found that no stability problems are encountered for $\mathscr{L} > 1$ and Fig. 7.11 is principally concerned to show the movement of the first two eigenvalues as $\mathscr{L}$ decreases below 1. The circled numbers are the values of $\mathscr{L}$ for which the eigenvalue is located at the nearest dot. When $\mathscr{L} = 1$ the first eigenvalue has come up from the origin to 3·124 and the second is to be found at 9·870. At $\mathscr{L} = 0{\cdot}9$ they have moved closer together and by $\mathscr{L} \doteqdot 0{\cdot}85$ they have coalesced on the real axis at 6. Thereafter they move as complex conjugates with decreasing real part until at $\mathscr{L} = 0{\cdot}4$ they are near the border of the stable half-plane. For a value of $\mathscr{L}$ only slightly less than 0·4 the steady state becomes unstable and remains so for smaller values of $\mathscr{L}$. At $\mathscr{L} = 0{\cdot}1$ both are again real—the first being $-62{\cdot}24$ and the second at $-5{\cdot}297$. During this process the larger eigenvalues move in the right-half plane sometimes on the real axis and some describing arcs in the complex plane. The first fourteen are distributed as follows:

Eigenvalues	$\mathscr{L}$	2·0	1·6	1·4	1·2	1·0	0·9	0·8	0·6	0·5	0·4	0·2	0·1
Real	Stable	12	12	10	14	14	2	4	6	8	8	10	12
	Unstable	—	—	—	—	—	—	—	—	—	—	—	2
Complex	Stable	2	2	4	0	0	12	10	8	6	6	2	—
	Unstable	—	—	—	—	—	—	—	—	—	—	2	—

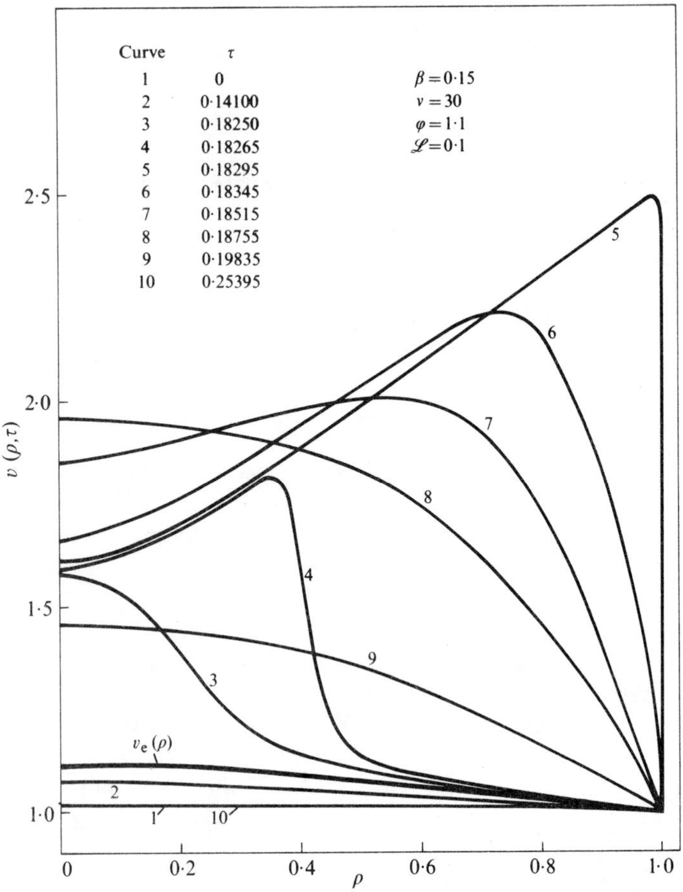

FIG. 7.12. The temperatures during the limit cycle around the unique unstable steady state (A of Fig. 7.10 when $\mathscr{L} = 0{\cdot}1$). Values of the dimensionless time of each curve are shown on the insert. (After Luss and Lee (1970).)

It should be remembered that there is a unique and unchanging steady state for all values of $\mathscr{L}$ and this analysis shows that it is unstable for values of $\mathscr{L}$ less than about 0·39. We thus have the phenomenon of a unique but unstable steady state and the invariant set may be a stable limit cycle. Figures 7.12 and 7.13 show that this is the case and illustrate the limit cycle as calculated by Luss and Lee. The unstable steady state is shown as the heavier line and the profiles are numbered for the sequence of values of τ shown in the figure. The system stays near the unstable steady state for some 80 per cent of the cycle (cf. curves 1, 2, 9, 10), but when a slight temperature rise develops at the centre of the sphere it propagates outwards to a hot spot (curve 5) in a rapidly expanding and receding wave. The time interval between curves 3 and 9 is only about 6 per cent of the period of the complete cycle.

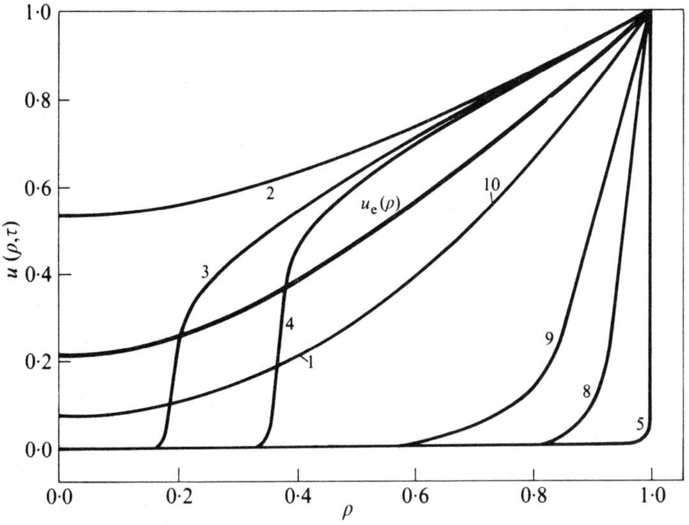

FIG. 7.13. The concentrations during the limit cycle of Fig. 7.12. The unstable steady state profile is shown as the heavy line in both figures. (After Luss and Lee (1970).)

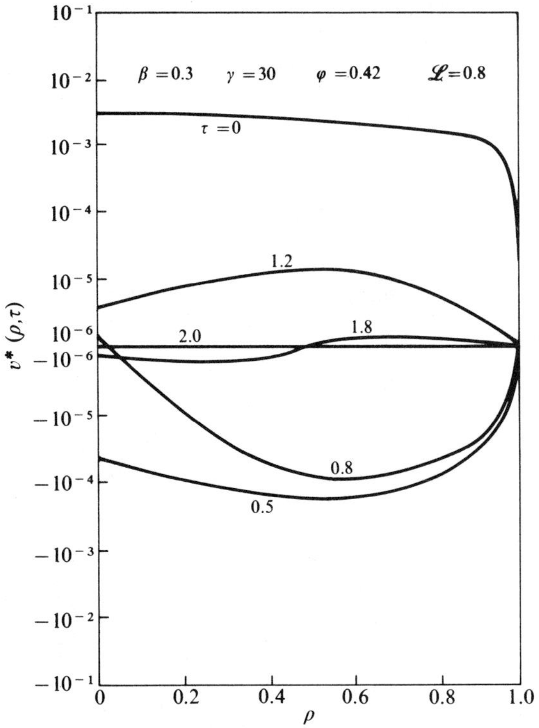

FIG. 7.14. The temperature deviation $v^*(\rho, \tau)$ in the approach to a stable high temperature steady state (case C_3 of Fig. 7.10) with $\mathscr{L} = 0{\cdot}8$. (After Luss and Lee (1970).)

The corresponding wave of concentration in Fig. 7.13 shows the reactant being rapidly exhausted as the high temperature wave propagates outwards. It is noticeable too how much this peak temperature exceeds the Prater temperature, in fact the temperature rise is some ten times greater than β, showing that the residual enthalpy can depart far from its equilibrium value when $\mathscr{L} < 1$. We shall revert to this point in the next chapter. Limit cycles of this type were first discovered by Hlaváček and Marek (1968*c*).

Further calculations were done by Luss and Lee for the points B, C_1, C_2, ... E on the curve for $\beta = 0{\cdot}3$ and they bring out some interesting differences for the case $\mathscr{L} = 1$. For example the analysis of Section 7.3 showed that for $\mathscr{L} = 1$ the approach to equilibrium was monotonic. By contrast when $\mathscr{L} = 0{\cdot}8$ the approach to the high temperature steady state C_3 showed considerable overshoot. Figure 7.14 shows the approach in this case represented on a logarithmic scale of temperature deviation v^*. The centre temperature first decreases falling below that of the steady state then increases, decreases and increases again before falling within the limits of $\pm 10^{-6}$ of steady state. By a time of $\tau = 2$ the departure from steady state is everywhere less than 10^{-6}. A non-monotonic approach to a steady state can also be obtained when $\mathscr{L} > 1$ as is shown in Fig. 7.15. The three heavy curves are the steady states corresponding to points C_1, C_2, and C_3 of Fig. 7.10 and the transient shown is $v_{1+\beta}(\rho, \tau)$. In contrast to the monotonic approach to the maximal solution shown by $v_{1+\beta}(\rho, \tau)$ in Fig. 7.5, the temperature at the centre

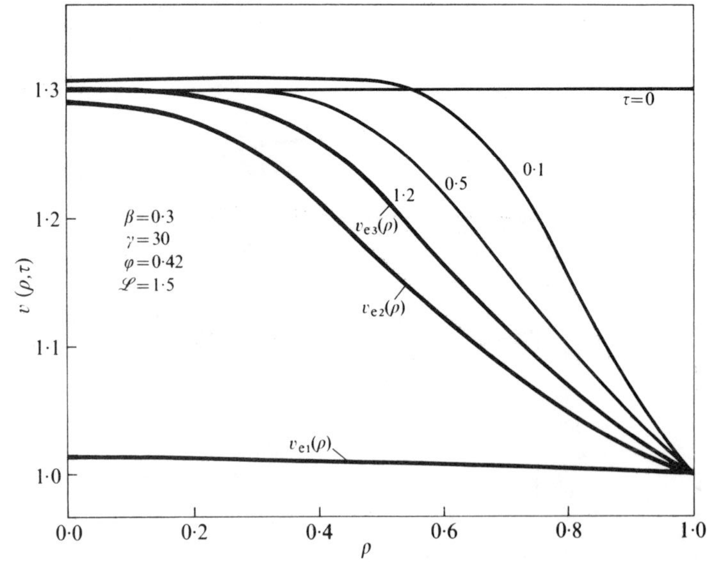

FIG. 7.15. The non-monotonic approach of $v_{1+\beta}(\rho, \tau)$ to the maximal steady state when $\mathscr{L} = 1{\cdot}5$. (After Luss and Lee (1970).)

of the sphere rises above the Prater temperature at first and then decreases to the maximal steady state value. The value of $\mathscr{L}$ may also affect the stable steady state to which the particular transient will tend. When $\mathscr{L} = 1$ the transient $v_{1+\beta}(\rho, \tau)$ always goes to the maximal steady state $v_{e3}(\rho)$, but Fig. 7.16 shows that this need not be so when $\mathscr{L} \neq 1$. For $\mathscr{L} = 0{\cdot}5$ the transient $v_{1+\beta}(\rho, \tau)$ is shown going to the minimal steady state $v_{e1}(\rho)$, and in fact

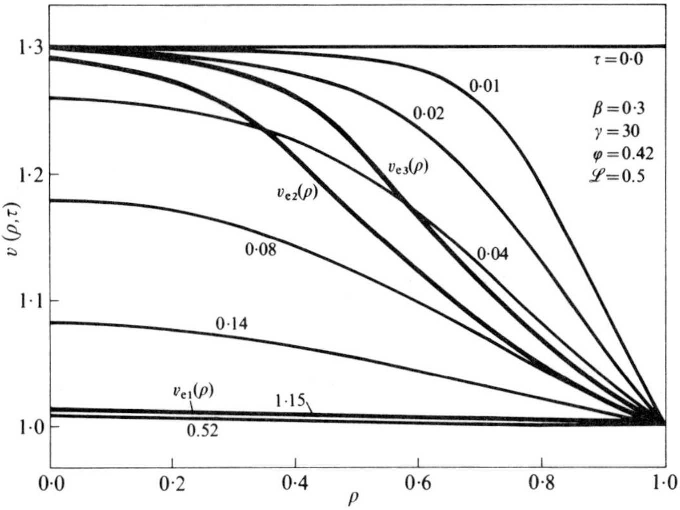

FIG. 7.16. The approach of $v_{1+\beta}(\rho, \tau)$ to the minimal steady state when $\mathscr{L} = 0{\cdot}5$.

overshooting it before becoming indistinguishably close at $\tau = 1{\cdot}15$. Thus for some value of $\mathscr{L}$ between 1 and 0·5 the initial condition $v_0(\rho) = 1+\beta$ has passed from the region of attraction of the maximal steady state to that of the minimal steady state.

These remarkable results of Luss and Lee show very clearly the variety of phenomena that the theory must save.

7.6. Linearization

The stability characteristics of the steady state have been examined in the previous section on the assumption that the local behaviour is adequately represented by neglecting squares and higher powers of the deviations from steady state. This is one of the classical approaches to stability problems and some instances of its use are worth citing.

Amundson and Raymond (1965) applied it to the isothermal catalyst pellet whose temperature might vary with time though not with position.

Thus

$$\frac{\partial u}{\mathrm{d}\tau} = \nabla^2 u - \phi^2 R(u, v_s) \quad \text{in } \Omega \tag{7.112}$$

$$u = u_s \quad \text{on } \partial\Omega \tag{7.113}$$

and v_s satisfies

$$\mathscr{L}\frac{\mathrm{d}v_s}{\mathrm{d}\tau} = \frac{\sigma\mu}{\upsilon}(1-v_s) + \frac{\beta\phi^2}{\upsilon}\iiint_\Omega R(u, v_s)\,\mathrm{d}\Upsilon. \tag{7.114}$$

The small deviations

$$u^*(\boldsymbol{\rho}, \tau) = (\boldsymbol{\rho}, \tau) - u_e(\boldsymbol{\rho}), \qquad v^*(\tau) = v(\tau) - v_{se} \tag{7.115}$$

satisfy

$$\frac{\partial u^*}{\partial \tau} = \nabla^2 u^* - \phi^2 R_u(u_e, v_{se})u^* - \phi^2 R_v(u_e, v_{se})v^*, \tag{7.116}$$

$$\mathscr{L}\frac{\mathrm{d}v^*}{\mathrm{d}\tau} = \frac{\beta\phi^2}{\upsilon}\iiint_\Omega R_u(u_e, v_{se})u^*\,\mathrm{d}\Upsilon + \left[\frac{\beta\phi^2}{\upsilon}\iiint_\Omega R_v(u_e, v_{se})\,\mathrm{d}\Upsilon - \frac{\sigma\mu}{\upsilon}\right]v^*, \tag{7.117}$$

with

$$u^* = 0 \quad \text{on } \partial\Omega, \tag{7.118}$$

and

$$u^*(\boldsymbol{\rho}, 0) = u_0^*(\boldsymbol{\rho}), \qquad v^*(0) = v_0^*. \tag{7.119}$$

Their technique was to take the Laplace transformation of these equations, giving equations for

$$\bar{u}^*(\boldsymbol{\rho}, s) = \int_0^\infty \mathrm{e}^{-s\tau}u^*(\boldsymbol{\rho}, \tau)\,\mathrm{d}\tau, \qquad \bar{v}^*(s) = \int_0^\infty \mathrm{e}^{-s\tau}v^*(\tau)\,\mathrm{d}\tau, \tag{7.120}$$

and to solve these by the use of a Green's function, showing later that the solutions will be asymptotically stable if a certain equation in the complex variable has roots only in the left half-plane. For a first-order irreversible reaction $R(u, v) = u \exp\{\gamma(v-1)/v\}$

$$\begin{aligned} R_u(u_e, v_{se}) &= E(v_{se}) = \exp\{\gamma(v_{se}-1)/v_{se}\}, \\ R_v(u_e, v_{se}) &= D(v_{se})u_e = (\gamma/v_{se}^2)E(v_{se})u_e, \end{aligned} \tag{7.121}$$

so that

$$s\bar{u}^* - u_0^*(\boldsymbol{\rho}) = \nabla^2\bar{u}^* - \phi^2 E(v_{se})\bar{u}^* - \phi^2 D(v_{se})u_e\bar{v}^*, \tag{7.122}$$

$$\mathscr{L}s\bar{v}^* - \mathscr{L}v_0^* = \frac{\beta\phi^2}{\upsilon}E(v_{se})\iiint_\Omega \bar{u}^*\,\mathrm{d}\Upsilon + \left[\frac{\beta\phi^2}{\upsilon}D(v_{se})\iiint_\Omega u_e\,\mathrm{d}\Upsilon - \frac{\sigma\mu}{\upsilon}\right]\bar{v}^*.$$

Equation (7.123) is solved for $\bar{v}^*$ treating $\iiint \bar{u}^*\,\mathrm{d}\Upsilon$ as a constant, υU, to be determined later. Thus

$$\bar{v}^* = \{\mathscr{L}v_0^* + \beta\phi^2 E(v_{se})U\}/\{\mathscr{L}s + (\sigma\mu/\upsilon) - \beta\phi^2 D\langle u_e\rangle\}, \tag{7.124}$$

where $\langle u_e\rangle$ is the average value of u_e over Ω. When this is substituted into eqn (7.122)

$$\nabla^2\bar{u}^* - (\phi^2 E + s)\bar{u}^* = (\phi^2 D\bar{v}^*)u_e(\boldsymbol{\rho}) - u_0^*(\boldsymbol{\rho}). \tag{7.125}$$

The right-hand side is a function of position with one coefficient depending linearly on U. Equation (7.125) can be solved using the Green's function for the operator $\{\nabla^2 - (\phi^2 E + s)\}$ and then averaged over Ω to provide a linear equation for U. Let $\mathscr{G}(\boldsymbol{\rho}, \boldsymbol{\rho}')$ be the Green's function and

$$U_e = \frac{1}{\upsilon}\iiint_\Omega \mathrm{d}\Upsilon \iiint_\Omega \mathscr{G}(\boldsymbol{\rho}, \boldsymbol{\rho}')u_e(\boldsymbol{\rho}')\,\mathrm{d}\Upsilon',$$

$$U_0 = \frac{1}{\upsilon}\iiint_\Omega \mathrm{d}\Upsilon \iiint_\Omega \mathscr{G}(\boldsymbol{\rho}, \boldsymbol{\rho}')u_0^*(\boldsymbol{\rho}')\,\mathrm{d}\Upsilon';$$

then

$$U = \frac{\mathscr{L}\phi^2 D v_0^* U_e - (\mathscr{L}s + (\sigma\mu/\upsilon) - \beta\phi^2 D\langle u_e\rangle)U_0}{\mathscr{L}s + (\sigma\mu/\upsilon) - \beta\phi^2 D\langle u_e\rangle - \beta\phi^4 DEU_e}. \tag{7.126}$$

The analysis then turns on the location of the zeros of the denominator. The first term is linear in s, the second and third are constant, but the third is a complicated function of s entering through the Green's function of $(\nabla^2 - \phi^2 E - s)$. For example, in the case of a slab the denominator is

$$\mathscr{L}s + \frac{\sigma\mu}{\upsilon} - \beta\phi^2 D\langle u_e\rangle - \beta\phi^4 DE\frac{1}{s}\left[\frac{\tanh(\phi^2 E + s)^{\frac{1}{2}}}{(\phi^2 E + s)^{\frac{1}{2}}} - \langle u_e\rangle\right]$$

and $\langle u_e\rangle = (\tanh \phi E^{\frac{1}{2}})/\phi E^{\frac{1}{2}}$. It need hardly be said that even for this simple case the analysis is more than a little intricate but, after some painful reduction, Amundson and Raymond (1965) arrive at necessary and sufficient condition for stability of the form

$$\frac{\sigma\mu}{\upsilon} \geqslant \frac{\beta\phi^2 D(v_{se})}{2}\left[\frac{\tanh \phi E^{\frac{1}{2}}}{\phi E^{\frac{1}{2}}} + \mathrm{sech}^2\,\phi E^{\frac{1}{2}}\right]. \tag{7.127}$$

In a later paper Amundson and Kuo (1967*c*) examine the case of the full equations for a single reaction; these can be linearized to the form

$$\frac{\partial w^*}{\partial \tau} = \mathbf{D}\nabla^2\mathbf{w}^* - \mathbf{C}\mathbf{w}^* \quad \text{in } \Omega,$$
$$\mathbf{D}\frac{\partial \mathbf{w}^*}{\partial \nu} + \mathbf{N}\mathbf{w}^* = 0 \qquad \text{on } \partial\Omega, \tag{7.128}$$

where

$$\mathbf{D} = \begin{bmatrix} 1 & \cdot \\ \cdot & 1/\mathscr{L} \end{bmatrix}, \quad \mathbf{C} = \begin{bmatrix} \phi^2 & \\ \cdot & -\beta\phi^2/\mathscr{L} \end{bmatrix}\begin{bmatrix} R_u & R_v \\ R_u & R_v \end{bmatrix}, \quad \mathbf{N} = \begin{bmatrix} \nu & \cdot \\ \cdot & \mathscr{L}\mu \end{bmatrix},$$

and the derivatives R_u and R_v are evaluated on the steady state under scrutiny. They show that though this problem is not self-adjoint it gives a sufficient condition for stability:

$$\|\mathbf{C}\| < \lambda_1^2, \tag{7.129}$$

where λ_1^2 is the least eigenvalue of the self-adjoint problem

$$\mathbf{D}\nabla^2\mathbf{w} + \lambda^2\mathbf{w} = 0 \quad \text{in } \Omega$$
$$\mathbf{D}\frac{\partial \mathbf{w}}{\partial \nu} + \mathbf{N}\mathbf{w} = 0 \quad \text{on } \partial\Omega. \tag{7.130}$$

A crude estimate of the norm of **C** can be obtained by applying Minkowski's inequality

$$\|\mathbf{C}\| \leqslant \phi^2(1+\beta^2/\mathscr{L}^2)^{\frac{1}{2}}\{\sup_{\Omega} R_u^2 + \sup_{\Omega} R_v^2\}^{\frac{1}{2}}. \tag{7.131}$$

This is very similar to the conditions obtained by the Liapounov function.

7.7. Lumping and other approximate methods

At the beginning of Chapter 6 we gave a plausible, if unrigorous, justification for the idea that the catalytic body and the stirred tank reactor would have a certain similarity of behaviour. This was used in Section 6.4.2 to show that facile bounds to the region of uniqueness could be obtained from the analogy. It can also be used to give some indications of the stability properties of the catalytic reaction though again it is necessary to be cautious in interpreting the results derived from such an analogous lumped system. They are best regarded as showing the likely trends but they must be handled with caution since quantitative agreement may not be more than fortuitous. More reliable conclusions may be drawn from better approximations, but it is worthwhile to see first what the crudest level of approximation has to offer.

The pair of ordinary differential equations describing a single reaction in the stirred tank is

$$\frac{\mathrm{d}u}{\mathrm{d}\tau'} = 1-u-\psi^2 R(u, v), \tag{7.132}$$

$$L\frac{\mathrm{d}v}{\mathrm{d}\tau'} = 1-v+\beta'\psi^2 R(u, v), \tag{7.133}$$

(cf. eqn (7.124)), and a steady state (u_e, v_e) is given by a pair of values of u and v such that

$$\begin{aligned} 0 &= 1-u_e-\psi^2 R(u_e, v_e), \\ 0 &= 1-v_e-\beta'\psi^2 R(u_e, v_e). \end{aligned} \tag{7.134}$$

Thus the departures from equilibrium

$$u^*(\tau') = u(\tau')-u_e, \qquad v^*(\tau') = v(\tau')-v_e \tag{7.135}$$

satisfy the linear equations

$$\frac{\mathrm{d}u^*}{\mathrm{d}\tau'} = -(1+\psi^2 R_u)u^* - \psi^2 R_v v^*, \tag{7.136}$$

$$L\frac{\mathrm{d}v^*}{\mathrm{d}\tau'} = \beta'\psi^2 R_u u^* - (1-\beta'\psi^2 R_v)v^* \tag{7.137}$$

when they are sufficiently small to have negligible squares and products. The derivatives R_u and R_v are constants for they are evaluated at (u_e, v_e). For this steady state to be asymptotically stable it is necessary and sufficient that

$$\left(1+\frac{1}{L}\right)+\psi^2\left(R_u-\frac{\beta'}{L}R_v\right) > 0 \tag{7.138a}$$

and

$$1+\psi^2(R_u-\beta' R_v) > 0. \tag{7.138b}$$

Using the interpretations of the parameters given in eqn (6.125) and considering for the moment the Dirichlet problem so that $\lambda_1(\mu) = \lambda_1(v) = \lambda_1$, the second of these conditions becomes

$$\phi^2\{\beta R_v - R_u\} < \lambda_1^2, \tag{7.139}$$

which is the condition we have repeatedly found; cf. eqns (7.84) and (7.70). The first condition, eqn (7.138a), which can be written

$$\phi^2 R_u(1-\mathscr{L}) < \mathscr{L}\lambda_1^2+\{\lambda_1^2-\phi^2(\beta R_v - R_u)\}, \tag{7.140}$$

is automatically satisfied if $\mathscr{L} \geqslant 1$ and eqn (7.139) is satisfied, but becomes more difficult to satisfy as $\mathscr{L}$ decreases. Only in the case of an endothermic reaction is there the possibility of satisfying both conditions for all values of $\mathscr{L}$ and ϕ. Again only eqn (7.138a) is necessary to ensure the uniqueness of the steady state (u_e, v_e) and if $\mathscr{L}$ is sufficiently small that the condition (7.140) is violated we shall have a unique unstable steady state.

Finlayson ((1972), Section 5.3, p. 123) has shown that this approximation of Hlaváček's is a one-term method of weighted residuals and that the one-term collocation method gives a similar result. Let u_1 and v_1 be the values of u and v at the collocation point and u_s, v_s the surface values. Then the one point collocation equations for the sphere

$$\begin{aligned}\frac{du_1}{d\tau} &= \frac{21}{2}(u_1 - u_s) - \phi^2 R(u_1, v_1),\\ \frac{dv_1}{d\tau} &= -\frac{21}{2\mathscr{L}}(v_1 - v_s) + \frac{\beta\phi^2}{\mathscr{L}} R(u_1, v_1),\end{aligned} \tag{7.141}$$

and the boundary conditions are

$$\tfrac{7}{2}(u_1 - u_s) = \nu(u_s - 1), \qquad \tfrac{7}{2}(v_1 - v_s) = \mu(v_s - 1). \tag{7.142}$$

The boundary conditions can be used to eliminate u_s and u_1 and then the linearized equations for the departures $u_1^* = u_1 + u_{1e}$, $v_1^* = v_1 - v_{1e}$ are

$$\begin{aligned}\frac{du_1^*}{d\tau} &= -(N + \phi^2 R_u)u_1^* - \phi^2 R_v v_1^*,\\ \frac{dv_1^*}{d\tau} &= \frac{\beta\phi^2}{\mathscr{L}} R_u u_1^* - \left(\frac{M}{\mathscr{L}} - \frac{\beta\phi^2}{\mathscr{L}} R_v\right) v_1^*,\end{aligned} \tag{7.143}$$

where

$$N = \frac{21\nu}{7 + 2\nu}, \qquad M = \frac{21\mu}{7 + 2\mu}. \tag{7.144}$$

This gives rise to the stability criteria

$$\begin{aligned}\phi^2\left\{\left(\frac{N}{M}\right)\beta R_v - R_u\right\} &< N,\\ \mathscr{L} &> (\beta R_v - M)/(N + \phi^2 R_u),\end{aligned} \tag{8.145}$$

and the second can be violated by making $\mathscr{L}$ sufficiently small if $\beta R_v > M$. As usual we notice that the criteria are always satisfied for endothermic reactions. Hellinckx, Grootjans, and van den Bosch (1972) have also remarked on this simple analogy and Padmanabhan and van den Bosch (1973) have considered it as the first of a hierarchy of collocation methods as mentioned by Finlayson.

We should consider two features of the general collocation procedure: the first a proof by Padmanabhan and van den Bosch (1974*b*) of the fact that as $\mathscr{L}$ decreases the eigenvalues always pass into the unstable half-plane as a pair of complex values; the second combination of this method with Liapounov functions by Perlmutter and McGowin (1971).

In Section 4.1.2 we saw that the *n*-point collocation methods developed by Villadsen and Stewart (1967) lead to replacing the spatial derivatives by weighted differences of the values at the collocation points. If this is done in the transient equations a set of coupled ordinary equations is obtained:

$$\frac{\mathrm{d}u_i}{\mathrm{d}\tau} = \sum_{i=1}^{n+1} B_{ij}u_j - \phi^2 R(u_i, v_i), \quad i = 1, \dots, n, \tag{7.146}$$

$$\mathscr{L}\frac{\mathrm{d}v_i}{\mathrm{d}\tau} = \sum_{j=1}^{n+1} B_{ij}v_j + \beta\phi^2 R(u_u, v_i), \quad (i = 1, \dots, n), \tag{7.147}$$

$$\nu = \sum_{j=1}^{n+1} A_j u_j + \nu u_{n+1}, \tag{7.148}$$

$$\mu = \sum_{j=1}^{n+1} A_j v_j + \mu v_{n+1}. \tag{7.149}$$

Here we have dropped the index (n) and $u_i = u(\rho_i, \tau)$, $v_i = v(\rho_i, \tau)$ where ρ_i is the position of the ith collocation point. ρ_{n+1} is on the surface and A_j is an abbreviation for $A_{n+1,j}$. The matrices for the symmetrical shapes are given by Villadsen and Stewart (1967) and Finlayson (1972). For other shapes the matrices could be calculated, but the formulae would be necessarily more complex for a number of different collocation points, not merely ρ_{n+1}, would lie on the surface.

If we consider the linearized equations for the deviations

$$u_i^*(\tau) = u(\rho_i, \tau) - u_e(\rho_i), \qquad v_i^*(\tau) = v(\rho_i, \tau) - v_e(\rho_i)$$

we have

$$\frac{\mathrm{d}u_i^*}{\mathrm{d}\tau} = \sum_{i=1}^{n} (N_{ij} - \phi^2 (R_u)_i \delta_{ij}) u_j^* - \phi^2 (R_v)_i v_i^*, \tag{7.150}$$

$$\frac{\mathrm{d}v_i^*}{\mathrm{d}\tau} = \frac{\beta\phi^2}{\mathscr{L}}(R_u)_i u_i^* + \sum_{i=1}^{n} \frac{1}{\mathscr{L}}\{M_{ij} + \beta\phi^2 (R_v)_i \delta_{ij}\} v_j^*, \tag{7.151}$$

where

$$N_{ij} = B_{ij} - \frac{A_j}{\nu + A_{n+1}}, \qquad M_{ij} = B_{ij} - \frac{A_j}{\mu + A_{n+1}}. \tag{7.152}$$

If $\mathbf{M}$ and $\mathbf{N}$ denote the matrices with elements M_{ij} and N_{ij} and

$$\mathbf{R}_u = \mathrm{diag}\{(R_u)_1, \dots (R_u)_n\}, \qquad \mathbf{R}_v = \mathrm{diag}\{(R_v)_1, \dots (R_v)_n\}$$

then the Jacobian matrix is the $2n \times 2n$ partitioned matrix

$$\mathbf{J} = \begin{bmatrix} \mathbf{N} - \phi^2 \mathbf{R}_u & -\phi^2 \mathbf{R}_v \\ \dfrac{\beta\phi^2}{\mathscr{L}} \mathbf{R}_u & \dfrac{1}{\mathscr{L}} \mathbf{M} + \dfrac{\beta\phi^2}{\mathscr{L}} \mathbf{R}_v \end{bmatrix}. \tag{7.153}$$

The eigenvalues for the n-point approximation will be the $2n$ roots of

$$|\mathbf{J} + \mu \mathbf{I}| = 0, \tag{7.154}$$

and the solution will be stable if the real parts of all the μ are positive. Padmanabhan and van den Bosch (1974*b*) point out that if any of the roots enter the unstable half-plane as the value of $\mathscr{L}$ is decreased it cannot be that the root which enters is real. For suppose $\mu = 0$ were a root, then we would have $|\mathbf{J}| = 0$. But because all the last n rows have a factor of $1/\mathscr{L}$, $|\mathbf{J}| = \mathscr{L}^{-n}|\mathbf{J}_1|$ where $\mathbf{J}_1$ is the matrix $\mathbf{J}$ with $\mathscr{L} = 1$. But this implies that $\mu = 0$ is an eigenvalue of $\mathbf{J}$ for all values of $\mathscr{L}$, which is contrary to the hypothesis that it was by changing $\mathscr{L}$ that the eigenvalue moved through the origin along the real axis. Since this argument is independent of the number of collocation points it suggests that it will also be true for the partial differential equations themselves. It *suggests*, but it does not prove, for to do this one would have to show that $\operatorname{Im} \mu > 0$ when $\operatorname{Re} \mu = 0$ and $n \to \infty$. The result is clearly analogous to the root locus in control theory.

Perlmutter and McGowin (1971) combine the collocation technique with that of the Liapounov function (see also Perlmutter, 1972). Thus if $\mathbf{w}^*$ denotes the column vector $(u_1^*, \dots u_n^*, v_1^*, \dots v_n^*)$, the linearized equations may be written

$$\frac{\mathrm{d}\mathbf{w}^*}{\mathrm{d}\tau} = \mathbf{J}\mathbf{w}^*. \tag{7.155}$$

If $\mathbf{K}$ is the matrix of eigenvectors of $\mathbf{J}$, so that

$$\mathbf{K}^{-1}\mathbf{J}\mathbf{K} = -\mathbf{L} = \operatorname{diag}(-\mu_1, -\mu_2, \dots -\mu_{2n}),$$

then

$$V = (\mathbf{w}^*)^T(\mathbf{K}^{-1})^T\mathbf{K}^{-1}\mathbf{w}^* = \mathbf{x}^T \cdot \mathbf{x} \tag{7.156}$$

is a suitable Liapounov function. For

$$\frac{1}{2}\frac{\mathrm{d}V}{\mathrm{d}\tau} = (\mathbf{w}^*)^T(\mathbf{K}^{-1})^T\mathbf{K}^{-1}\mathbf{J}\mathbf{w}^* = -\mathbf{x}^T\mathbf{L}\mathbf{x} \tag{7.157}$$

so that if all the μ_i have positive real parts the positive definite V has a negative definite derivative. The full nonlinear equations

$$\frac{\mathrm{d}\mathbf{w}^*}{\mathrm{d}\tau} = \mathbf{S}\mathbf{w}^* + \mathbf{R}(\mathbf{w}^*) \tag{7.158}$$

where

$$\mathbf{S} = \begin{bmatrix} \mathbf{N} & \mathbf{0} \\ \mathbf{0} & \mathbf{M}/\mathscr{L} \end{bmatrix}; \qquad \mathbf{R} = \begin{bmatrix} \operatorname{diag} R^*(u_i^*, v_i^*) & \mathbf{0} \\ \mathbf{0} & \mathscr{L}^{-1} \operatorname{diag} R^*(u_i^*, v_i^*) \end{bmatrix} \tag{7.159}$$

and

$$R^*(u_i^*, v_i^*) = R(u_{ie} + u_i^*, v_{ie} + v_i^*) - R(u_{ie}, v_{ie}),$$

are then used in calculating the derivative of V, giving

$$\frac{\mathrm{d}V}{\mathrm{d}\tau} = \dot{V} = 2(\mathbf{w}^*)^T(\mathbf{K}^{-1})^T\mathbf{K}^{-1}\{\mathbf{S}\mathbf{w}^* + \mathbf{R}(\mathbf{w}^*)\}. \tag{7.160}$$

To find the region of asymptotic stability Perlmutter and McGowin used a method developed by Perlmutter in earlier studies which gives the largest ellipsoid $V =$ constant for which $\dot{V} < 0$. Points on the boundary of this ellipsoid may be converted into profiles of concentration and temperature by the interpolating Jacobi polynomials, and so map out the regions of asymptotic stability in the state space. For the case $\mathscr{L} = 1$ using only three collocation points, they were able to get regions of asymptotic stability covering very large fractions of the known regions of attraction of the stable maximal and minimal solutions. For $\mathscr{L} \neq 1$ the regions of asymptotic stability found by this method formed comparatively narrow bounds around the steady state solutions. The variation of these regions with the number of collocation points shows that this method suffers from the unpredictability common to all these approximations.

Luss and Lee (1971) use yet another lumping technique. Consider the equations for the concentrations of the species taking part in the reaction $\Sigma\alpha_i A_i = 0$, namely

$$\frac{\partial u_i}{\partial \tau} = \Delta_i \nabla^2 u_i + \alpha_i \phi^2 R(u_1, \ldots u_S), \tag{7.161}$$

and let

$$\langle u_i \rangle = \frac{1}{\upsilon} \iiint_{\Omega} u_i(\boldsymbol{\rho}, \tau)\, \mathrm{d}\Upsilon \tag{7.162}$$

be the average concentration of species A_i. Internal and overall mass transfer coefficients, κ_i and K_i, are then defined by

$$\kappa_i(u_{is} - \langle u_i \rangle) = K_i(1 - \langle u_i \rangle) = \frac{\Delta_i}{\upsilon} \iint_{\partial\Omega} \frac{\partial u_i}{\partial \nu}\, \mathrm{d}\Sigma \tag{7.163}$$

so that with the Robin boundary conditions

$$\frac{1}{K_i} = \frac{1}{\kappa_i} + \frac{1}{\nu_i}. \tag{7.164}$$

Then

$$\frac{\Delta_i}{\upsilon} \iiint\limits_{\Omega} \nabla^2 u_i \, \mathrm{d}\Upsilon = \frac{\sigma}{\upsilon}\Delta_i \iint\limits_{\partial\Omega} \frac{\partial u_i}{\partial \nu} \, \mathrm{d}\Sigma = \frac{\sigma}{\upsilon} K_i (1 - \langle u_i \rangle),$$

and averaging and linearizing the equations gives

$$\begin{aligned}\frac{\mathrm{d}\langle u_i^* \rangle}{\mathrm{d}\tau} &= \frac{\sigma}{\upsilon} K_{ie} \langle u_i^* \rangle + \alpha_i \phi^2 \sum_{j=1}^{S} \langle (R_{u_j})_e u_j^* \rangle \\ &= \frac{\sigma}{\upsilon} K_{ie} \langle u_i^* \rangle + \alpha_i \phi^2 \sum_{j=1}^{S} \{\langle (R_{u_j})_e \rangle \langle u_j^* \rangle + \langle M_j u_j^* \rangle\},\end{aligned} \tag{7.165}$$

where

$$M_j = (R_{u_j})_e - \langle (R_{u_j})_e \rangle. \tag{7.166}$$

The lumping method now makes two main assumptions. The first is that asymptotic stability on the average, namely that the $\langle u_i^* \rangle$ all tend to zero as $\tau \to \infty$, implies asymptotic stability of the steady state. The second is that the M_j may be neglected. The first assumption is plausible from what we know of the generally 'reasonable' behaviour of parabolic equations. The second can be checked to see how much variation there is with the M_j and if it is too much the region Ω may be broken up into a number of subregions in each of which the variation is sufficiently small. Luss and Lee (1971) show how this can be done, but also show that in several cases they considered comparatively high values of $M_j/\langle (R_{u_j})_e \rangle$ did not disturb the general correctness of their conclusions. However, as is always the case with lumpings and approximations, the validity of the results can only be established by a meta-investigation and accuracy in one problem may be lost in another.

Another method of lumping is to restrict attention to models in which one or more of the variables is held constant. Such was the approach of Amundson and Kuo (1967*a*, *b*) and it has been used by others mentioned in the bibliographical notes later. McGreavy and Thornton (1970*b*) have applied it to the situation where the pellet is always at a spatially constant temperature though this may vary in time. They assume in addition that the steady state distribution of concentration is instantaneously taken up. Thus for the single irreversible reaction

$$v(\boldsymbol{\rho}, \tau) = v_s(\tau), \tag{7.167}$$

$$\nabla^2 u - \phi^2 R(u, v_s) = 0, \tag{7.168}$$

$$\left(\frac{\partial u}{\partial \nu}\right) + \nu u_s = \nu. \tag{7.169}$$

The isothermal problem defined by (7.168) and (7.169) can be solved to give an effectiveness factor

$$\eta(v_s; \nu, \phi) = \frac{1}{\upsilon} \iiint_{\Omega} R(u, v_s)\, d\Upsilon. \tag{7.170}$$

Then averaging the equation for $v(\boldsymbol{\rho}, \tau)$ and using eqn (7.167) and the boundary condition for v gives

$$\mathscr{L}\frac{\partial v_s}{\partial \tau} = \frac{\sigma}{\upsilon}\mu(1-v_s) + \beta\phi^2\eta(v_s; \nu, \phi) \tag{7.171}$$

which is a nonlinear ordinary differential equation for the temperature of the particle. For the slab, for example,

$$\eta(v_s; \nu, \phi) = \left\{\frac{\phi^2 e^2(v_s)}{\nu} + \frac{\phi e(v_s)}{\tanh \phi e(v_s)}\right\}^{-1} \tag{7.172}$$

where $e^2(v_s)$ is an abbreviation for $\exp(\gamma - \gamma/v_s)$. McGreavy and Thornton study the system under sinusoidal fluctuations of external temperature. In

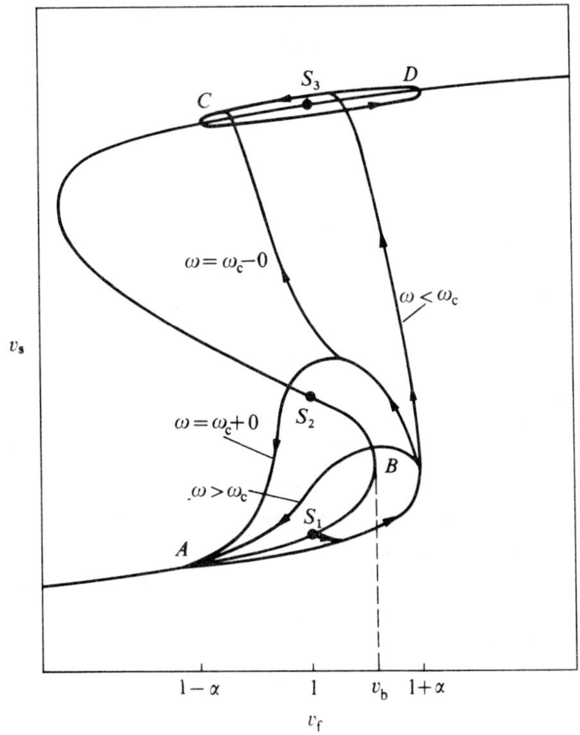

FIG. 7.17. Forced oscillations of an isothermal catalyst particle in the plane of v_s, the catalyst temperature, and v_f, the external temperature: $v_f = 1 + \alpha \sin \omega t$.

our notation this would be done by replacing the 1, which represents the external temperature, in eqn (7.171) by $1+\alpha \sin \omega t = v_f$. These authors show that though the external temperature may pass into the region of attraction of a very high-temperature steady state, this does not always result in a temperature runaway. Figure 7.17 is a qualitative sketch of the kind of result which they were able to obtain. For the mean external temperature $v_f = 1$ there are three steady state values of v_s shown by the points S_1, S_2, and S_3 on the steady state (v_s, v_f)-curve. The amplitude α is such that $v_f = 1+\alpha$ is beyond v_b, the bifurcation value of v_f, so that if the variation of v_f were very slow the low temperature steady state on the branch AS_1B would only obtain until v_f reached v_b. For a slight increase beyond v_b the temperature of the pellet would rise rapidly to the high temperature branch and the point (v_f, v_s) would continue to move slowly backwards and forwards on the arc CS_3D of the high-temperature branch. However if the frequency ω is sufficiently large, the excursus of v_f above the bifurcation value v_b is sufficiently short that the subsequent decrease in temperature of the surroundings brings the catalyst temperature down in a forced cycle near the branch AB. There is, however, a critical frequency and if $\omega < \omega_c$, the excursus above v_b lasts long enough for the state to be attracted to the upper branch, and its subsequent movement is a narrow cycle around the arc CS_3D. A further study of the critical frequency and amplitude at the limit of stability has been made by McGreavy and Soliman (1973).

7.8. Finite amplitude methods

Linearization gives results on local stability because it deals essentially in terms of infinitesimal disturbances and it is only by the retention of the non-linear terms as in Liapounov's direct method, that estimates of the region of asymptotic stability can be obtained. For the stability problems of fluid mechanics methods have been developed which allow the estimation of the amplitudes of disturbances which are critical either for stability or instability and these have been very neatly put together and applied to the catalyst problem by Denn (1973). The combination of a variational method with the Liapounov functional allows an estimate of the size of disturbance which we can be sure is stable: this, in the fluid mechanical context stems from the work of Serrin. On the other hand by continuing the expansion beyond the linear terms an estimate of the amplitude of a disturbance which is sure to be unstable can be made, by the methods of Stuart and Eckhaus. As Denn has pointed out their relationship may be most clearly seen in a simple model problem.

Consider

$$\frac{\partial w}{\partial \tau} = \frac{\partial^2 w}{\partial \rho^2} + aw(1+\tfrac{1}{2}bw) \tag{7.173}$$

in $0 < \rho < 1$ with

$$\frac{\partial w}{\partial \rho} = 0, \quad \rho = 0, \qquad w = 0, \quad \rho = 1,$$

where a and b are constants. The linearized problem is satisfied by a solution of the form

$$w(\rho, \tau) = \sum c_n w_n(\rho) \exp(-\lambda_n^2 \tau) \tag{7.174}$$

if

$$w_n'' + (a + \lambda_n^2) w_n = 0 \tag{7.175}$$

subject to $w_n'(0) = w_n(1) = 0$. This means that

$$\lambda_n^2 = (2n-1)^2 \frac{\pi^2}{4} - a, \quad w_n(\rho) = 2^{\frac{1}{2}} \cos(2n-1)\frac{\pi\rho}{2}.$$

Consider now the Liapounov functional

$$V(\tau) = \int_0^1 \tfrac{1}{2} w^2 \, \mathrm{d}\rho \tag{7.176}$$

for which, as before,

$$-V'(\tau) = \int_0^1 \left\{ \left(\frac{\partial w}{\partial \rho} \right)^2 - a(1 + \tfrac{1}{2}bw)w^2 \right\} \mathrm{d}\rho. \tag{7.177}$$

Stability will be assured if

$$a < \lambda^2 = \min \left\{ \int_0^1 \left(\frac{\partial w}{\partial \rho} \right)^2 \mathrm{d}\rho \right\} \Big/ \left\{ \int_0^1 (1 + \tfrac{1}{2}bw)w^2 \, \mathrm{d}\rho \right\}, \tag{7.178}$$

and in seeking this minimum any trial functions can be used which satisfy the boundary conditions. The Euler equation for this problem is

$$w'' + \lambda^2 w(1 + \tfrac{3}{4}bw) = 0 \tag{7.179}$$

with $w'(0) = w(1) = 0$, and for this a regular perturbation solution can be sought. Thus, suppose

$$w = Aw_1,$$

where

$$A^2 = \int_0^1 w^2(\rho) \, \mathrm{d}\rho$$

is a measure of the amplitude of the disturbance, then substitution in eqn (7.179) gives

$$-\lambda_1^2 A w_1 + \lambda^2 A w_1 (1 + \tfrac{3}{4} b A w_1) = 0.$$

Multiplying by $w_1(\rho)$ and integrating gives

$$\lambda^2 = \lambda_1^2\left\{1+\left(\tfrac{3}{4}b\int_0^1 w_1^3\,\mathrm{d}\rho\right)A\right\}^{-1}$$

since w_1 is normalized. The stability condition (7.178) can then be rearranged to give

$$A < A_c = \tfrac{4}{3}\frac{\lambda_1^2 - a}{|ab\int_0^1 w_1^3\,\mathrm{d}\rho|} = \frac{\pi(\pi^2-4a)}{8\cdot 2^{\frac{1}{2}}|ab|} \tag{7.180}$$

and A_c is a critical amplitude below which we can be sure of stability.

In Eckhaus's formulation a solution of the form

$$\sum_{n=1}^{\infty} A_n(\tau)w_n(\rho) \tag{7.181}$$

is sought and an infinite set of equations for the A_n is obtained by substituting in eqn (7.173), multiplying by $w_n(\rho)$ and integrating. Since the $w_n(\rho)$ are orthonormal this reduces to

$$\frac{\mathrm{d}A_n}{\mathrm{d}\tau} = -\lambda_n^2 A_n + \tfrac{1}{2}ab\sum_{l=1}^{\infty}\sum_{m=1}^{\infty} c_{lmn}A_l A_m, \tag{7.182}$$

where

$$c_{lmn} = \int_0^1 w_l(\rho)w_m(\rho)w_n(\rho)\,\mathrm{d}\rho. \tag{7.183}$$

The amplitude functions are now transformed to bring all terms to the order of unity by letting

$$A_n(\tau) = \frac{2\lambda_1^4}{a\lambda_n^2}a_n(\tau),$$

so that

$$\frac{\mathrm{d}a_n}{\mathrm{d}\tau} = -\lambda_n^2 a_n + b\lambda_n^2\sum\sum c_{lmn}\frac{\lambda_1^4}{\lambda_l^2\lambda_m^2}a_l a_m.$$

The critical value of a for neutral stability is $a = \frac{1}{4}\pi^2$ since then $\lambda_1 = 0$. In this limit the only terms which survive in the double sum are those with $l = m = 1$ for all others become small as λ_1^4. Hence a solution near this neutral point will have

$$\frac{\mathrm{d}a_n}{\mathrm{d}\tau} = -\lambda_n^2 a_n + b_n^2 c_{11n}a_1^2,$$

or

$$\frac{\mathrm{d}A_n}{\mathrm{d}\tau} = -\lambda_n^2 A_n + \tfrac{1}{2}abc_{11n}A_1^2.$$

The idea is that the disturbance is initially dominated by the first eigenfunction of eqn (7.175) and hence A_1 acts as a forcing function in all the other Fourier coefficients. If it is unbounded then all the other coefficients will be unbounded and hence the condition for A_1 to grow without bound gives an estimate of the amplitude of disturbance which is sure to be unstable. But $A_1(\tau)$ satisfies

$$\frac{\mathrm{d}A_1}{\mathrm{d}\tau} = -\lambda_1^2 A_1 + \tfrac{1}{2}abc_{111}A_1^2 \tag{7.184}$$

and so is

$$A_1(\tau) = A_1(0)\exp(-\lambda_1^2\tau)\Big/\left[1 - A_1(0)\frac{abc_{111}}{2\lambda_1^2}\left\{1-\exp(-\lambda_1^2\tau)\right\}\right] \tag{7.185}$$

If $A_1(0) > |2\lambda_1^2/abc_{111}|$ then $A_1(\tau) \geqslant A_1(0)$ grows without bound. Since $\|w\|^2 \geqslant A_1^2(\tau)$ the disturbance will grow without bound if

$$A \geqslant A^{\mathrm{c}} = |a\lambda_1^2/abc_{111}|. \tag{7.186}$$

Thus these methods combine to give an upper bound A_{c} for the amplitude of disturbances which decay, eqn (7.180) and a lower bound A^{c} for the amplitude of those that certainly grow, eqn (7.186). In this particularly simple case Denn remarks that $A_{\mathrm{c}}/A^{\mathrm{c}} = \frac{2}{3}$ so that we may hope for a fairly good coverage of the state space.

In applying this to the catalyst particle we will consider only the simple case of $\mathscr{L} = 1$ and adiabatic perturbations ($z^* = 0$) so that we may use the single equation (7.48)

$$\frac{\partial v^*}{\partial \tau} = \nabla^2 v^* + \beta\phi^2\mathscr{F}(v_{\mathrm{e}}, v^*)v^*$$

in Ω with

$$v^* = 0 \quad \text{on } \partial\Omega.$$

Denn has shown how to apply it to the general case, though he remarks that the algebraic labours of the Eckhaus method are painful. As before an appropriate Liapounov function is

$$V(\tau) = \tfrac{1}{2}\iiint_{\Omega} [v^*(\boldsymbol{\rho}, \tau)]^2 \,\mathrm{d}\Upsilon, \tag{7.187}$$

and by the usual manipulations (cf. Section 7.4)

$$-V'(\tau) = \iiint_{\Omega} \{(\nabla v^*)^2 - \beta\phi^2\mathscr{F}(v_{\mathrm{e}}, v^*)(v^*)^2\}\,\mathrm{d}\Upsilon.$$

If

$$\lambda^2 = \min\left\{\iiint_\Omega (\nabla w)^2 \, \mathrm{d}\Upsilon\right\} \Big/ \left\{\beta \iiint_\Omega \mathscr{F}(v_\mathrm{e}, w)(w)^2 \, \mathrm{d}\Upsilon\right\}, \quad (7.188)$$

the minimum being over all functions for which $w = 0$ on $\partial\Omega$, then stability will be assured if $\phi < \lambda$. But the minimizing function satisfies the Euler equation

$$\nabla^2 w + \lambda^2 \beta \mathscr{F}(v_\mathrm{e}, w) w + \tfrac{1}{2}\lambda^2 \beta \mathscr{F}'(v_\mathrm{e}, w) w^2 = 0 \quad (7.189)$$

in Ω with $w = 0$ on $\partial\Omega$, where

$$\mathscr{F}'(v_\mathrm{e}, w) = \frac{\partial}{\partial w}\mathscr{F}(v_\mathrm{e}, w). \quad (7.190)$$

If $w_1(\boldsymbol{\rho})$ and λ_1^2 are the first eigenfunction and eigenvalue of the linearized problem

$$\nabla^2 w + \lambda^2 \beta \mathscr{F}(v_\mathrm{e}, 0) w = 0 \quad (7.191)$$

then the procedure outlined with the model equation leads to the estimate

$$\lambda^2 = \lambda_1^2\left[1 - \frac{3}{4}\left\{\frac{\iiint \mathscr{F}(v_\mathrm{e}, 0) w_1^3 \, \mathrm{d}\Upsilon}{\iiint \mathscr{F}(v_\mathrm{e}, 0) w_1^2 \, \mathrm{d}\Upsilon}\right\} A + O(A^2)\right].$$

Now, from eqn (7.50), $\mathscr{F}(v_\mathrm{e}, 0) = F'(v_\mathrm{e})$ and $\mathscr{F}'(v_\mathrm{e}, 0) = F''(v_\mathrm{e})$, so that the estimate of the maximum stable amplitude is

$$A_\mathrm{c} = \frac{4}{3}\left\{1 - \frac{\phi^2}{\lambda_1^2}\right\}\left|\frac{\iiint F'(v_\mathrm{e}) w_1^2 \, \mathrm{d}\Upsilon}{\iiint F''(v_\mathrm{e}) w_1^3 \, \mathrm{d}\Upsilon}\right|. \quad (7.192)$$

In the Eckhaus formalism of the problem of finding A^c the method follows that of the model problem. The $w_n(\boldsymbol{\rho})$ and λ_n^2 are the normalized eigenfunctions and eigenvalues of the linearized equation (7.19) and the $A_n(\tau)$ satisfies

$$\frac{\mathrm{d}A_n}{\mathrm{d}\tau} = -\lambda_n^2 A_n + \tfrac{1}{2}\beta\phi^2 \sum_l \sum_m c_{lmn} A_l A_m,$$

where

$$c_{lmn} = \iiint \{F''(v_\mathrm{e}(\boldsymbol{\rho})) w_l(\boldsymbol{\rho}) w_m(\boldsymbol{\rho}) w_n(\boldsymbol{\rho})\} \, \mathrm{d}\Upsilon.$$

With this definition we again have

$$A^\mathrm{c} = |2\lambda_1^2/\beta\phi^2 c_{111}| \quad (7.193)$$

and the ratio $(A_\mathrm{c}/A^\mathrm{c})$ is of the order of $\frac{2}{3}$.

7.9. The catalytic wire

In Section 4.7.3 we derived the equations for the temperature and concentration of a catalytic wire in the steady state and discussed the periodic solutions. The transient equations for $u(\zeta, \tau)$ and $v(\zeta, \tau)$, the concentration and temperature at position ζ and time τ, are

$$\frac{\partial u}{\partial \tau} = v'(1-u) - R(u, v), \tag{7.194}$$

$$L\frac{\partial v}{\partial \tau} = \frac{\partial^2 v}{\partial \zeta^2} + \mu'(1-v) + R(u, v). \tag{7.195}$$

The parameters and variables are defined in eqns (4.369), and the additional parameter introduced in the transient equations is the heat capacity term L, so designated for its analogous role to the Lewis number.

$$L = C_{\mathrm{p}} A T_{\mathrm{f}}/(-\Delta H) p c_{\mathrm{f}}, \tag{7.196}$$

where A is the area and p the perimeter of a cross-section of the wire and C_{p} its heat capacity per unit volume. The wire is assumed to be so fine that radial gradients are unimportant.

If the longitudinal conduction of heat is ignored or only solutions independent of ζ are entertained, the equations are simply those of the stirred tank and have been fully treated as such, both theoretically and experimentally by Luss and Cardoso (1969). In a later study Luss and Ervin (1972*b*) discussed the flickering due to the fluctuations of velocity of turbulent flow past the wire. These will affect the mass and heat transfer coefficients which are taken to be proportional to a power of the Reynolds number. Thus for the case of solutions independent of ζ we would have ordinary differential equations

$$\begin{aligned} \frac{\mathrm{d}u}{\mathrm{d}\tau} &= v' f(\tau)(1-u) - R(u, v), \\ L\frac{\mathrm{d}v}{\mathrm{d}\tau} &= \mu' f(\tau)(1-v) + R(u, v), \end{aligned} \tag{7.197}$$

where $f(\tau)$ is some function fluctuating about a mean value of 1. To simplify matters Luss and Ervin take this to be a sinusoidal fluctuation of amplitude α and period $2\pi/\omega$,

$$f(\tau) = (1 + \alpha \sin \omega\tau)^n. \tag{7.197a}$$

The eigenvalues associated with the equations linearized about a steady state $u_{\mathrm{e}}, v_{\mathrm{e}}$ can be calculated as the roots of the quadratic

$$\mu^2 - \left(v' + R_u + \frac{\mu'}{L} - \frac{1}{L}R_v\right)\mu + \frac{1}{L}\left(\mu' v' + \frac{\mu'}{L}R_u - \frac{v'}{L}R_v\right) = 0.$$

When they are complex, the imaginary part $\bar{\omega}$, is a natural resonant frequency of the system, and it might be expected that fluctuations of this frequency would excite a considerable variation of temperature in the non-linear equations (7.197). This is borne out in Fig. 7.18 which is based on

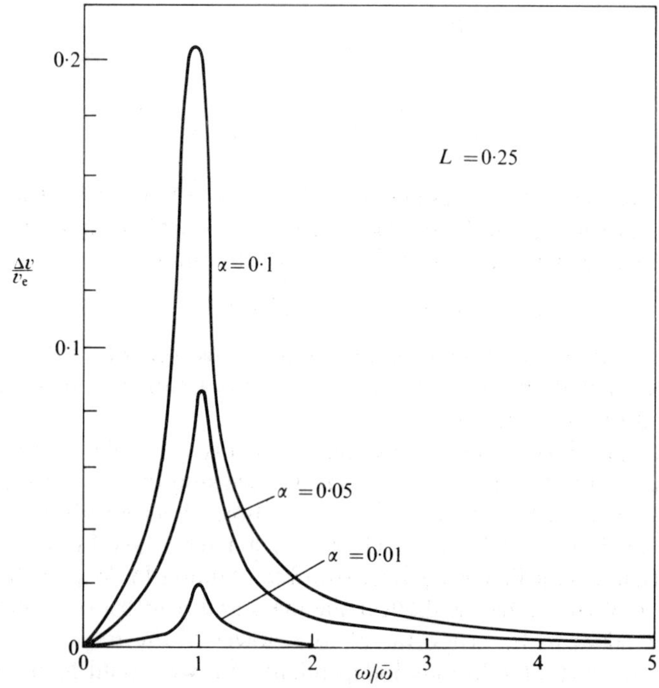

FIG. 7.18. The amplitude of the temperature response, Δv, as a fraction of the mean temperature, v_e, and a function of α, the fractional amplitude of the perturbation, and ω, its frequency. (After Luss and Ervin.)

Luss and Ervin's calculations. It shows the amplitude of the temperature fluctuations as a fraction of the steady state temperature as a function of $\omega/\bar{\omega}$. The response peaks to an amplitude of the order of the amplitude of the velocity fluctuations, namely 2α, in the neighbourhood of $\omega = \bar{\omega}$. These calculations were done with an exponent $n = 0{\cdot}38$ and with $L = 0{\cdot}25$. At higher values of L the response is considerably less for the heat capacity of the system damps the oscillations. If the eigenvalues of the linearized system at the steady state about which fluctuations are imposed are both real, then the amplitude of the response can be two orders of magnitude smaller than that of the excitation. There is clearly some affinity between this type of analysis and the studies of McGreavy and Thornton (1970*b*) mentioned above. Luss, Worley and Edwards (1973) have recently reported experiments which indicate a correlation between turbulent velocity fluctuations and flickering.

Luss and Ervin (1972*a*) also looked at the influence of end conditions on a finite length of wire. The concentration and temperature will satisfy eqns (7.194) and (7.195) subject to the boundary condition (in their case) of

$$v = 1, \qquad \zeta = 0, Z. \tag{7.198}$$

The steady state solutions may be found by solving

$$v'(1-u_e) = R(u_e, v_e) \tag{7.198a}$$

for u as a function of v and substituting in

$$\frac{d^2 v_e}{d\zeta^2} = \mu'(v_e - 1) - R(u_e, v_e). \tag{7.199}$$

As we saw in Section 5.7.3, the case of a first-order reaction leads to the pair of first-order equations

$$\begin{aligned} \frac{dv_e}{d\zeta} &= w_e \\ \frac{dw_e}{d\zeta} &= \mu'(v_e - 1) - \frac{v' \exp(\gamma - \gamma/v_e)}{v' + \exp(\gamma - \gamma/v_e)} = g(v_e) \end{aligned} \tag{7.200}$$

Note that the only difference between these equations and eqn (4.374) is the suffix e which is necessary here to distinguish the steady state. The trajectories in the phase plane must pass between points on the line $v_e = 1$ and be such that the integral of $\{w_e(v_e)\}^{-1}$ with respect to v_e is

$$Z = l\{(-\Delta H)pc_f\, e^{-\gamma}/kA\}^{\frac{1}{2}}$$

where l is the length of the wire.

The critical points of the phase plane are where $w_e = 0$ and $g(v_e) = 0$. The latter equation will have one or three roots according to the values of μ', v', and γ. This can be determined by Fig. 6.21 taking $\alpha = e^{\gamma}/v'$ for the index of the curve, $(\gamma\mu'/v')$ as abscissa and $1/\gamma$ as ordinate. Figure 7.19 shows the phase plane when there is one critical point, C. This is a saddle-point through which the trajectories DD′ and EE′ pass. The trajectory representing a solution will be just such an arc as AB on which

$$\int_A^B \frac{dv_e}{w(v_e)} = Z. \tag{7.201}$$

It is clear that it will always be possible to find such a solution for the integral tends to zero as the arc AB approaches O, whereas it tends to infinity as AB gets near the curve DCE. The point C itself corresponds to the uniform solution in a very long wire for which the influence of the end conditions

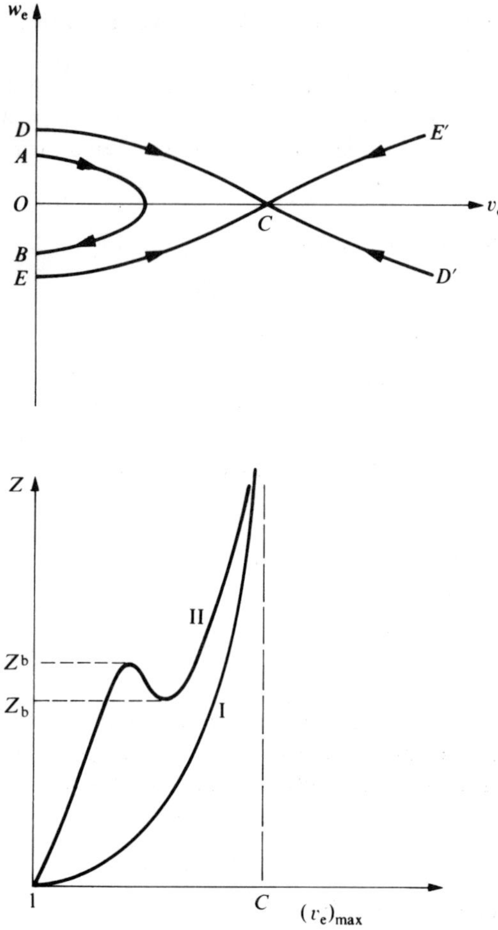

FIG. 7.19. The phase plane for the catalytic wire with only one uniform state.

is negligible. It follows that the maximum temperature in any non-uniform solution is less than that of the uniform solution for a long wire. If Z, as calculated from eqn (7.201), is plotted against $(v_e)_{max}$, the point at which the trajectory crosses the v_e-axis a curve such as is illustrated in the lower part of the figure is obtained. It may be like the curve (I) and give a unique steady state or like (II) and yield three solutions for a range of lengths $Z_b < Z < Z^b$.

When the function $g(v_e)$ is as shown in the upper part of Fig. 7.20 there are three possible uniform steady states, v_1, v_2, and v_3 for an infinitely long wire. When the area between the axis and the g-curve from v_1 to v_2 is greater than the area below the axis from v_2 to v_3 the phase plane is as

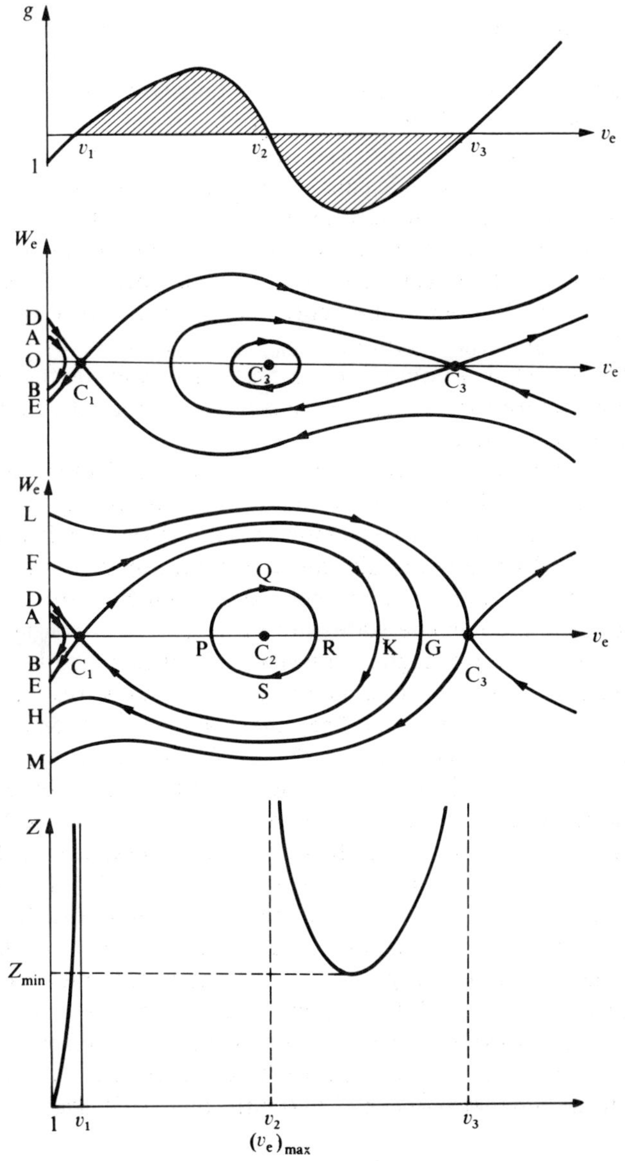

FIG. 7.20. The phase plane for the catalytic wire when three uniform steady states are possible.

shown in the next section of the figure. The non-uniform solutions are much the same as in Fig. 7.19, namely arcs AB lying in the segment DC_1E. It follows that the maximum temperature is less than the lowest uniform temperature C_1. If however the area under the g-curve from v_1 to v_2 is less than the area below the axis from v_2 to v_3, then the phase plane is as shown in the third section of the figure. Then not only arcs like AB can connect two points on the boundary $v_e = 1$ but there can also be arcs like FGH for which the maximum temperature lies between v_2 and v_3. As the points F and H move toward L and M respectively the arc gets closer to LC_3M and the value of Z calculated from eqn (7.101) approaches infinity. Similarly as F and H move toward D and E respectively the arc approaches DC_1KC_1E and again Z increases without bound due to the contribution to Z of the part of the integral near C_1. Thus the locus of Z versus $(v_e)_{max}$ is as shown at the bottom of the figure and has one branch running from $(v_e)_{max} = 1$, $Z = 0$ to the asymptote $(v_e)_{max} = v_1$ generated by arcs like AB, and another branch depending from the vertical asymptotes through v_2 and v_3. It follows that for wires longer than a certain Z_{min} there are three possible non-uniform steady states.

Luss and Ervin encountered extreme numerical difficulties in analysing the stability of the solutions and were forced to do experimental transient calculations. By this means they demonstrated that the stability of a steady state was markedly affected by the capacity parameter L and that the high temperature steady state could be rendered unstable by a sufficiently small value of L. The intermediate state with $dZ/d(v_e)_{max} < 0$ can be shown to be unstable by the type of argument given in Amundson and Luss (1967*a*).

Luss and Ervin (1972*a*) also allude to the standing waves on an infinite wire or wire with insulated ends. They show, as does Jackson (1972*a*), that such solutions are always unstable. These solutions have been derived in Section 4.7.3, and correspond to the closed paths in the phase planes of Fig. 7.20 which surround the unstable uniform state C_2 (cf. Fig. 5.46 or the curve PQRS in Fig. 7.21). If $2Z$ is the wave length, an extremum of the wave will correspond to P or R, and the integral of eqn (7.201) calculated round PQRSP is $2Z$. Let the origin correspond to a maximum such as R, then $\zeta = Z$ will be a minimum such as P and $g(v_e(0)) < 0$, $g(v_e(z)) > 0$. Jackson considers the case of very large L for which the concentration response is much quicker than that of the temperature. In this case $\tau' = \tau/L$ can be taken as the dimensionless time and the factor of $(1/L)$ in front of $\partial u/\partial \tau'$ in eqn (7.194) means that we can neglect the derivative of u. The substituting from the steady state form of eqn (7.194) into eqn (7.195) gives

$$\frac{\partial v}{\partial \tau'} = \frac{\partial^2 v}{\partial \zeta^2} - g(v) \tag{7.202}$$

where $g(v)$ is the same function as in eqn (7.200). It follows that the linearized equation for the deviation $v^*(\zeta, \tau') = v(\zeta, \tau') - v_e(\zeta)$ will be

$$\frac{\partial v^*}{\partial \tau'} = \frac{\partial^2 v^*}{\partial \zeta^2} - g'(v_e(\zeta))v^*, \tag{7.203}$$

with

$$\frac{\partial v^*}{\partial \zeta} = 0 \quad \text{at} \quad \zeta = 0 \quad \text{and} \quad \zeta = Z. \tag{7.204}$$

A solution of separable form $V(\zeta)\,\mathrm{e}^{-\sigma\tau'}$ will be stable if the real parts of the eigenvalues σ are positive, and $V(\zeta)$ will satisfy

$$\frac{\mathrm{d}^2 V}{\mathrm{d}\zeta^2} - \{g'(v_\mathrm{e}(\zeta)) - \sigma\}V = 0$$

with

$$\frac{\mathrm{d}V}{\mathrm{d}\zeta} = 0, \quad \zeta = 0, Z. \tag{7.205}$$

But Amundson (1965) shows that $\mathrm{Re}\,\sigma$ is not negative if the solution of the equation

$$\frac{\mathrm{d}^2 U}{\mathrm{d}\zeta^2} - g'(v_\mathrm{e}(\zeta))U = 0,$$
$$U(0) = 1, \qquad \left(\frac{\mathrm{d}U}{\mathrm{d}\zeta}\right)_{\zeta=0} = 0 \tag{7.206}$$

is strictly positive throughout $(0, Z)$ and

$$\frac{1}{U}\frac{\mathrm{d}U}{\mathrm{d}\zeta} \geqslant 0 \quad \text{at } \zeta = Z.$$

But differentiating the second of eqns (7.200) with respect to ζ shows that w_e satisfies

$$\frac{\mathrm{d}^2 w_\mathrm{e}}{\mathrm{d}\zeta^2} - g'(v_\mathrm{e}(\zeta))w_\mathrm{e} = 0 \tag{7.207}$$

with $w_\mathrm{e}(0) = 0$, $(\mathrm{d}w_\mathrm{e}/\mathrm{d}\zeta)_0 = g(v_\mathrm{e}(0))$. Multiplying eqn (7.206) by w_e and eqn (7.207) by U, subtracting and integrating from $\zeta = 0$ to $\zeta = Z$ gives

$$w_\mathrm{e}(0)U'(0) - U(0)w_\mathrm{e}'(0) = w_\mathrm{e}(Z)U'(Z) - U(Z)w_\mathrm{e}'(Z).$$

Using the boundary conditions $w_\mathrm{e}(0) = 0$, $w_\mathrm{e}'(0) = g(v_\mathrm{e}(0))$, $U(0) = 1$, $U'(0) = 0$ gives

$$g\{v_\mathrm{e}(0)\} = U(Z)g\{v_\mathrm{e}(Z)\}. \tag{7.208}$$

But we have seen that $g\{v_\mathrm{e}(0)\} < 0$ and $g\{v_\mathrm{e}(Z)\} > 0$, and it follows that $U(Z) < 0$ and the condition for stability is violated.

Thus the class of solutions on the catalytic wire which lack the symmetry of the boundary conditions through not being invariant under translations is unstable. We turn now to the stability problem for other unsymmetrical solutions.

7.10. The stability of unsymmetrical solutions

We wish in this section to sketch the stability analysis of the unsymmetrical solutions of Marek (Section 3.6.3) and Pismen and Kharkats (Section 4.7.1). This (to adapt a phrase from Titchmarsh's Preface to *The Theory of Fourier Integrals*) has been 'done as analysts should' by Aronson and Peletier (1974), and we shall do no more than outline their elegant results. They used, and refined for their purpose, a development of the Westphal–Prodi theorem made by Peterson and Maple (1966) which deserves mention in its own right.

Consider a very general differential equation in one space variable ρ and the time τ,

$$u_\tau = F(\rho, u, u_\rho, u_{\rho\rho}). \tag{7.209}$$

We can always normalize the interval of interest to be $-1 \leqslant \rho \leqslant 1$ and will use suffixes of $+$ and $-$ to denote quantities appropriate to the two boundaries. The boundary conditions are equally general, namely;

$$\begin{aligned} u_\rho(-1, \tau) &= f_-(u(-1, \tau)), \\ u_\rho(1, \tau) &= f_+(u(1, \tau)). \end{aligned} \tag{7.210}$$

We assume that f_+ and f_- are continuous with bounded first derivatives and that F is a continuously differentiable function of its arguments and is non-decreasing in $u_{\rho\rho}$. A solution of eqns (7.209) and (7.210) for which

$$u(\rho, 0) = u_0(\rho) \tag{7.211}$$

will be denoted by $u(\rho, \tau; u_0)$. The initial data is assumed to be twice continuously differentiable. A steady state solution $u_e(\rho)$ is such that $u(\rho, \tau; u_e) = u_e$. If it is asymptotically stable its region of asymptotic stability is

$$A = \{(\rho, u); -1 \leqslant \rho \leqslant 1, \psi_*(\rho) < u < \psi^*(\rho)\} \tag{7.212}$$

such that, if $u_0 \in A$, then

$$\lim_{\tau \to \infty} \sup_{\rho \in [-1, 1]} |u(\rho, \tau; u_0) - u_e(\rho)| = 0.$$

The theorem of Peterson and Maple envisages a monotonically increasing one-parameter family of solutions of the steady state equation

$$F(\rho, v, v_\rho, v_{\rho\rho}) = 0 \tag{7.213}$$

which we will denote by $v(\rho; \lambda)$, $\lambda_* \leqslant \lambda \leqslant \lambda^*$, and such that $u_e(\rho) = v(\rho; \tilde{\lambda})$ for some $\tilde{\lambda}$ in the open interval (λ_*, λ^*). In addition this one-parameter family fails to satisfy the boundary conditions in different ways according as λ is in $(\lambda_*, \tilde{\lambda})$ or in $(\tilde{\lambda}, \lambda^*)$. This can be illustrated best by the two figures

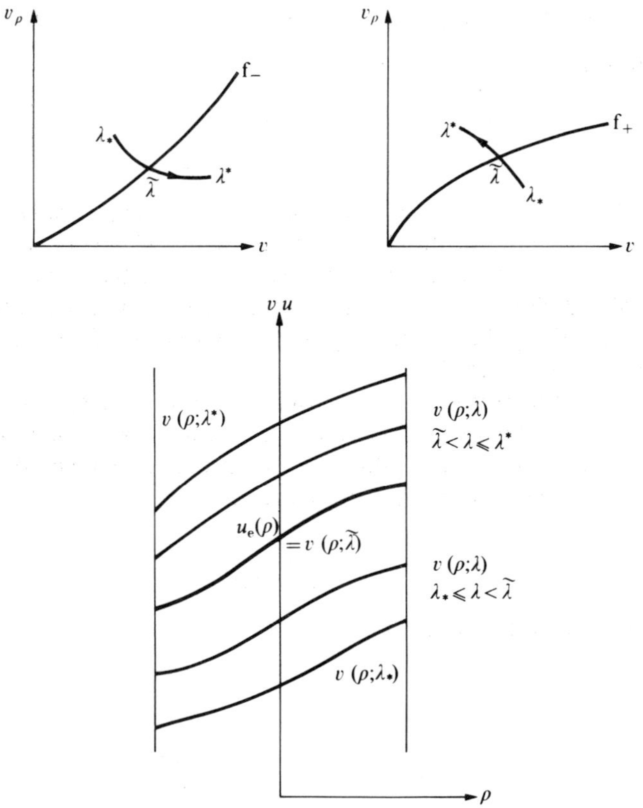

FIG. 7.21. Illustrating the Peterson–Maple stability criterion.

associated with the two boundaries at the top of Fig. 7.21. The curves of $f_+(v)$ and $f_-(v)$ are sketched in their respective (v_ρ, v)-planes, and of course at $\lambda = \tilde{\lambda}$ the points $(v_-, v_{\rho-})$ and $(v_+, v_{\rho+})$ fall exactly on the curves f_+ and f_- respectively. However, as λ is increased from λ_* to λ^* the points $(v(-1;\lambda), v_\rho(-1))$ and $(v(1;\lambda), v_\rho(1;\lambda))$ move across the respective curves in opposite directions. With this as background we may state the theorem more formally. Let $v(\rho;\lambda)$ be a one-parameter family of solutions of the ordinary differential equation (7.213) with the properties that:

(a) there is a $\tilde{\lambda}$ in (λ_*, λ^*) such that $v_\rho(-1;\tilde{\lambda}) = f_-(v(-1;\tilde{\lambda}))$ and $v_\rho(1;\lambda) = f_+(v(1;\tilde{\lambda}))$;
(b) $v_\lambda(\rho;\lambda) > 0$, $-1 \leqslant \rho \leqslant 1$, $\lambda_* \leqslant \lambda \leqslant \lambda^*$;
(c_*) $v_\rho(-1;\lambda) > f_-(v(-1;\lambda))$, $v_\rho(1;\lambda) < f_+(v(1;\lambda))$, $\lambda_* \leqslant \lambda < \tilde{\lambda}$;
(c^*) $v_\rho(-1;\lambda) < f_-(v(-1;\lambda))$, $v_\rho(1;\lambda) > f_+(v(1;\lambda))$, $\tilde{\lambda} < \lambda \leqslant \lambda^*$.

Then $u(\rho, \tau; v(\rho;\tilde{\lambda})) = u_e(\rho)$ is a steady state solution of eqns (7.209), (7.210).

From above $F_{v_{\rho\rho}} \geqslant 0$ and if, in addition,

(d) either $$F_v \neq 0$$

or $$F_v = 0, F_{v_\rho} \neq 0 \tag{7.214}$$

or $$F_v = F_{v_\rho} = 0, F_{v_{\rho\rho}} \neq 0$$

for $-1 \leqslant \rho \leqslant 1$, $\lambda_* \leqslant \lambda \leqslant \lambda^*$, then $u_e(\rho)$ is asymptotically stable with a region of asymptotic stability given by eqn (7.212) with $\psi_*(\rho) = v(\rho;\lambda_*)$, $\psi^*(\rho) = v(\rho;\lambda^*)$. If conditions (a), (b) and (d) hold but both the inequalities are violated either in (c_*) or in (c^*), then the steady state is unstable. The lower part of Fig. 7.21 shows the relative disposition of the various solutions that have been referred to.

Now this theorem can be very readily adapted to the situation of surface reaction on a slab for the solutions of the steady state equations are always straight lines. Let us recall that Marek's problem involved the diffusion of a reactant across a slab of porous material,

$$\frac{\partial u}{\partial \tau} = \frac{\partial^2 u}{\partial \rho^2}, \tag{7.215}$$

with reaction at the surfaces $\rho = \pm 1$,

$$\begin{aligned} 1-u_- &= R(u_-) - 2v\left(\frac{\partial u}{\partial \rho}\right)_-, \\ 1-u_+ &= R(u_+) + 2v\left(\frac{\partial u}{\partial \rho}\right)_+. \end{aligned} \tag{7.216}$$

Since in the steady state

$$u(\rho) = \tfrac{1}{2}(u_+ + u_-) + \tfrac{1}{2}(u_+ - u_-)\rho, \tag{7.217}$$

the values of u_+ and u_- must satisfy

$$u_+ = F(u_-), \qquad u_- = F(u_+), \tag{7.218}$$

where

$$F(u) = F(u;v) = \left(1+\frac{1}{v}\right)u - \frac{1}{v} + \frac{1}{v}R(u). \tag{7.219}$$

As an alternative to the graphical construction of the solution given in Section 3.6.3, the method shown in Fig. 7.22 has an advantage; it was in fact the method used by Marek and by Pismen and Kharkats. In the plane of u_+ and u_- the two curves $u_+ = F(u_-)$ and $u_- = F(u_+)$ are drawn. Their intersections will give possible pairs of values of u_+ and u_-. If $1-u-R(u) = 0$ has three solutions, then for sufficiently small v the curves will be disposed as shown in Fig. 7.22. There are three symmetrical steady states and six

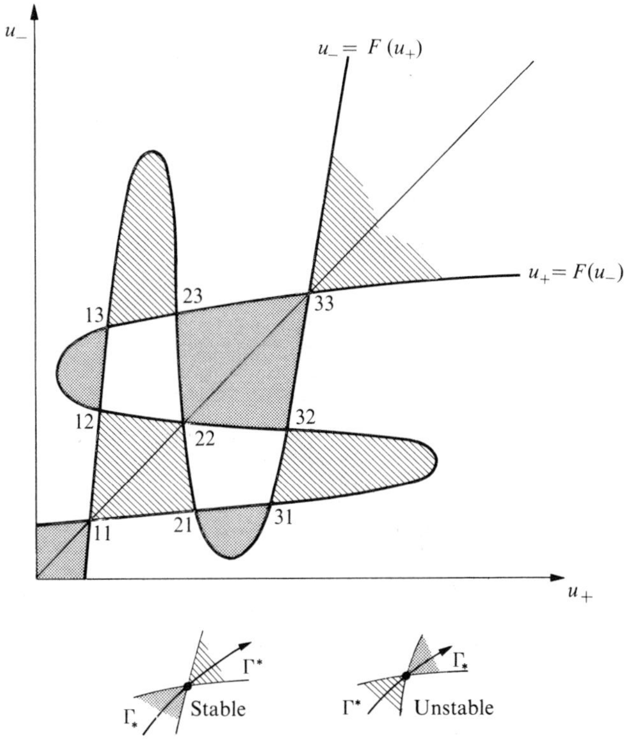

FIG. 7.22. Steady states in the (u_+, u_-)-plane and stability.

unsymmetrical ones which are numbered to correspond to those shown in Fig. 3.40.

Now any solution of the steady state equation

$$\frac{d^2v}{d\rho^2} = 0$$

is linear in ρ and so a one-parameter family of solutions is

$$v(\rho;\lambda) = \tfrac{1}{2}\{v_+(\lambda)+v_-(\lambda)\}+\tfrac{1}{2}\{v_+(\lambda)-v_-(\lambda)\}\rho. \qquad (7.220)$$

Each solution corresponds to a point in the (v_+, v_-)-plane and the family is represented by a segment of a curve parametrized by λ. If we superpose the (v_+, v_-)-plane on the (u_+, u_-)-plane, we can take this curve segment to go through one of the steady states, and the value of λ to be $\tilde{\lambda}$ at the steady state. We then have just such a family as is referred to in the theorem and have satisfied condition (a). Condition (b) will be satisfied if we take $v_+(\lambda)$ and $v_-(\lambda)$ to be increasing functions of λ, i.e. we consider curve segments that have a tangent in the direction of increasing λ which lies in the first

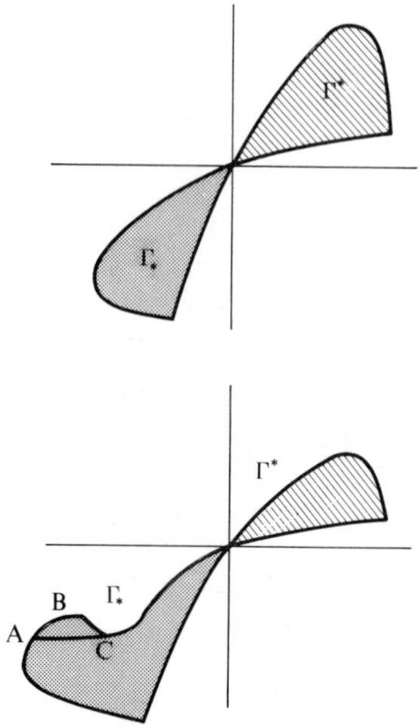

FIG. 7.23. Two types of stable steady state.

quadrant. We now distinguish two types of region in the plane: the region shown stippled, for which $u_+ > F(u_-)$, $u_- > F(u_+)$, and those shown cross-hatched, for which $u_+ < F(u_-)$ and $u_- < F(u_+)$. Let these types of region be called respectively Γ_* and Γ^*. Then inspection of eqns (7.216) and (7.220) shows that condition (c_*) is satisfied in any region of type Γ_* and (c^*) is satisfied in any region of type Γ^*. The condition (d) of the stability theorem is automatically satisfied. Thus the stability theorem can be expressed in geometrical terms as follows. An intersection of the two curves in Fig. 7.22 represents a stable steady state it if is possible to pass through it *from* a Γ_* region *to* a Γ^* region along a curve whose positive tangent lies in the first quadrant. On this basis we see immediately that the symmetrical states 11 and 33 and the unsymmetrical states 13 and 31 are stable.

More than this can be learned from such a diagram for the Peterson–Maple theorem also gives information about the region of asymptotic stability. The disposition of the curves in the neighbourhood of a steady state is shown enlarged in Fig. 7.23. When the Γ_* and Γ^* regions lie wholly in the third and first quadrants respectively and are such that any vertical or horizontal line intersects them in only one interval, then any point of the Γ_* region may be used for $v(\rho;\lambda_*)$ and be connected to any $v(\rho;\lambda^*)$ corresponding to a point of the Γ^* region. It follows that the region of asymptotic stability includes the area between the envelope of all possible lines represented by points of the two regions. If the Γ_* region were as shown in the lower part of the figure, then the part ABC would not be accessible to legitimate members of the family $v(\rho;\lambda)$.

This is the essence of Aronson and Peletier's work, though the reader will find their results and a sharpened form of the Peterson–Maple theorem stated with the proper completeness and precision. The method is clearly applicable to both the isothermal exterior problem of Marek (Section 3.6.2) and the non-isothermal problem of Pismen and Kharkats (Section 4.7.1). In conjunction with a maximum principle it might also serve for the distributed problem of Jackson and Horn where the boundary values of a subsidiary Dirichlet problem might play the role of u_+ and u_-.

7.11. The stability theory of chemotaxis and aggregation

The cellular slime mould (*Acrasiales*) exhibits at a certain stage of its life cycle a tendency to depart from a uniform distribution and form clumps or aggregates. It is known that this is mediated by acrasin which is degraded by an enzyme acrasinase in the medium outside the cells. This phenomenon, one of the classic instances of chemotaxis, has been the subject of a stability analysis by Segel and Keller (1970*a*, *b*). They consider a two-dimensional medium so that the dependent variables are functions of two spatial co-ordinates ρ_1 and ρ_2 and the time τ. Let v denote the concentration of cells and u that of acrasin which is produced by the cells at a rate $f(u)v$. Its degradation by the acrasinase is at a rate $g(u)$; Keller and Segel took this to be governed by Michaelis–Menten kinetics, but it may be left completely general. A mass balance on the acrasin concentration therefore gives

$$\frac{\partial u}{\partial \tau} = \nabla^2 u - g(u) + f(u)v. \qquad (7.221)$$

The movement of the cells is by their own diffusion, with the usual Fickian flux $-\Delta_1 \nabla v$, and chemotactically in the direction of increasing acrasin concentration with a flux $\Delta_2 \nabla u$. Thus v satisfies the equation

$$\frac{\partial v}{\partial \tau} = \nabla . (\Delta_1 \nabla v) - \nabla . (\Delta_2 \nabla u), \qquad (7.222)$$

since Δ_1 and Δ_2 could be functions of u and v. There is a uniform steady state (u_e, v_e) where

$$v_e = g(u_e)/f(u_e), \qquad (7.223)$$

and Keller and Segel raise the question of whether this is stable.

The linearized equations for small departures from uniformity are

$$\frac{\partial u^*}{\partial \tau} = \nabla^2 u^* - \{g'(u_e) - f'(u_e)v_e\}u^* + f(u_e)v^*, \tag{7.224}$$

$$\frac{\partial v^*}{\partial \tau} = -\Delta_2(u_e, v_e)\nabla^2 u^* + \Delta_1(u_e, v_e)\nabla^2 v^*. \tag{7.225}$$

There are no boundary conditions on this simplified problem, it being assumed that the medium is effectively infinite in extent. If solutions

$$\begin{aligned} u^* &= \tilde{u}\cos(s_1\rho_1 + s_2\rho_2)\,\mathrm{e}^{-\sigma\tau}, \\ v^* &= \tilde{v}\cos(s_1\rho_1 + s_2\rho_2)\,\mathrm{e}^{-\sigma\tau} \end{aligned} \tag{7.226}$$

are sought and s^2 denotes $s_1^2 + s_2^2$, the $\tilde{u}$ and $\tilde{v}$ will have to satisfy

$$\begin{aligned} \{\sigma - s^2 - g'(u_e) + f'(u_e)v_e\}\tilde{u} + f(u_e)\tilde{v} &= 0, \\ s^2\Delta_2\tilde{u} + (\sigma - s^2\Delta_1)\tilde{v} &= 0. \end{aligned} \tag{7.227}$$

The condition that these shall have non-trivial solutions gives a quadratic

$$\sigma^2 - \sigma\{s^2(1+\Delta_1) + g'(u_e) - f'(u_e)v_e\} + s^2\{s^2\Delta_1 + \Delta_1 g'(u_e) - \Delta_1 f'(u_e)v_e - \\ - \Delta_2 f(u_e)\} = 0. \tag{7.228}$$

The condition that a disturbance of wave number s should decay is that the roots of this quadratic, should be positive: they can easily be shown to be real. Thus for stability we must have

$$s^2(1+\Delta_1) > f'(u_e)v_e - g'(u_e) \tag{7.229}$$

and

$$s^2 > \frac{\Delta_2}{\Delta_1} f(u_e) + f'(u_e)v_e - g'(u_e). \tag{7.230}$$

The fastest growing instability corresponds to $s = 0$, so that a sufficient condition for stability of all disturbances will be

$$g'(u_e) - f'(u_e)v_e > \frac{\Delta_2}{\Delta_1} f(u_e) \tag{7.231}$$

since this makes both the right-hand sides of eqns (7.229) and (7.230) negative. This condition can be written as

$$\frac{\mathrm{d}u_e}{\mathrm{d}v_e} < \frac{\Delta_1}{\Delta_2}, \tag{7.232}$$

where u_e is defined as a function of v_e by the relation (7.223). Thus if the equilibrium acrasin concentration level does not increase too rapidly with increasing concentration of cells, the uniform equilibrium solution will be stable. The ratio Δ_1/Δ_2 gives a precise measure of what we mean by not increasing too rapidly. If the mobility of the cells is large (Δ_1 large) or the chemotactic effect small (Δ_2 small) then the tolerable value of du_e/dv_e is larger. Keller and Segel (1970*a*, *b*) discuss this in terms of the competition of factors which make for the growth of the disturbance with those that make for its decay. In this case the former is changed by an amount $\Delta_2\, du_e$ whilst the latter is controlled by $\Delta_1\, dv_e$. So long as $\Delta_1\, dv_e > \Delta_2\, du_e$ we have a stable situation. Disturbances of the form given by eqn (7.226) would result in clumping along parallel lines, and a pattern of such disturbances might lead to patterned clumping as with the hexagonal pattern given by the combination

$$\{\cos s\rho_1 + \cos(\tfrac{1}{2}s\rho_1 - \tfrac{1}{2}\sqrt{3}s\rho_2) + \cos(\tfrac{1}{2}s\rho_1 + \tfrac{1}{2}\sqrt{3}s\rho_2)\}\, e^{-\sigma\tau}.$$

If the system is confined to a circular area of radius $\rho = (\rho_1^2 + \rho_2^2)^{\frac{1}{2}} = P$ and we look for radially symmetric disturbances a suitable form of disturbance would be

$$u^* = \tilde{u}J_0(j_1\rho/P)\, e^{-\sigma\tau}, \qquad v^* = \tilde{v}J_0(j_1\rho/P)\, e^{-\sigma\tau},$$

where j_1 is the first zero of the derivative of J_0, for then $\partial u^*/\partial\rho$ and $\partial v^*/\partial\rho$ vanish for $\rho = P$. The stability analysis follows the same lines of substituting these expressions in the equations and finding the condition that σ should be positive. Again the determinantal condition is sufficient and gives

$$\frac{du_e}{dv_e} < \frac{\Delta_1}{\Delta_2}\left(1 - \frac{\Delta_1 j_1^2}{\Delta_2 P^2 f(u_e)}\right)^{-1}. \tag{7.233}$$

Because the geometry restricts the wave number of the disturbance that may be formed the system is more stable in the sense that (du_e/dv_e) has to be larger for instability. We see that as $P \to \infty$ the condition reduces to eqn (7.232) as it should, but more interestingly there is a critical size of circle such that the uniform state is always stable for amoebae confined to smaller circles. Using the numerical value $j_1 = 3{\cdot}8$ and writing $A_c = \pi P_c^2$ for the area of this critical circle

$$A_c = 45\Delta_1/\Delta_2 f(u_e) \tag{7.234}$$

or

$$\Delta_2[A_c f(u_e)v_e] = 45\Delta_1 v_e.$$

The term in the square bracket is the total production rate of acrasin in the circle and if this is too small no aggregation will occur. In any case the smaller the circle the larger must be the gradient (du_e/dv_e) before instability arises, and delayed aggregation in small drops is a known phenomenon.

If the uniform state is unstable we may ask what the form of the non-uniform steady state may be. This will be dictated by the geometry of the confining region and will be the solution of

$$\nabla^2 u = g(u) - f(u)v, \tag{7.235}$$

$$\nabla \cdot (\Delta_1 \nabla v - \Delta_2 \nabla u) = 0. \tag{7.236}$$

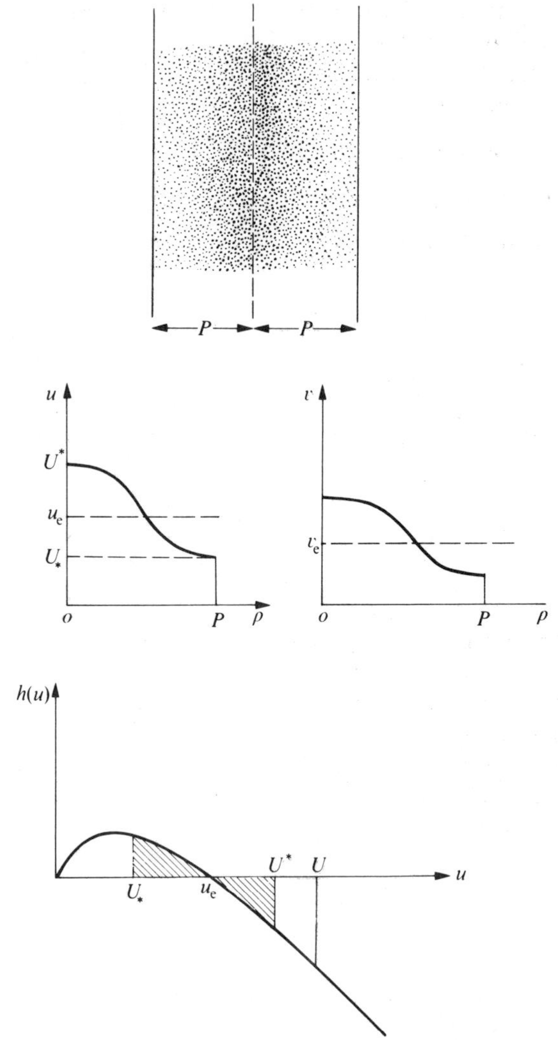

FIG. 7.24. The stable aggregation formed along the centre line of a long dish under chemotaxis.

We will consider only the simplest case of constant Δ_1 and Δ_2 and suppose that a disturbance has induced a clumping around the centre line of a long rectangular dish of breadth $2P$, as shown in Fig. 7.24. Then the Laplacian is $\mathrm{d}^2/\mathrm{d}\rho^2$, where ρ is the distance from the centre line, and the boundary conditions are the vanishing of all derivatives at $\rho = 0$ and $\rho = P$. Integration of eqn (7.236) then gives immediately that

$$\Delta_1 v - \Delta_2 u = \text{constant} = \Delta_2 c,$$

where

$$c = (\Delta_1 v_e - \Delta_2 u_e)/\Delta_2 = \{\Delta_1 g(u_e) - \Delta_2 u_e f(u_e)\}/\Delta_2 f(u_e).$$

Let

$$h(u) = g(u) - \frac{\Delta_2}{\Delta_1}(u+c)f(u), \tag{7.237}$$

which vanishes for $u = 0$ and $u = u_e$ and because we are supposing the uniform state to be unstable,

$$h'(u_e) < 0. \tag{7.238}$$

The concentration $u(\rho)$ will vary from a maximum of U^* at the centre to a minimum U_* at $\rho = P$, with an inflection point where $u = u_e$ (see centre section of Fig. 7.24). By integrating

$$\frac{\mathrm{d}^2 u}{\mathrm{d}\rho^2} = h(u) \tag{7.239}$$

by the usual quadratures we see that U_* and U^* are such that

$$\int_{U_*}^{u_e} h(u)\,\mathrm{d}u = \int_{u_e}^{U^*} -h(u)\,\mathrm{d}u \tag{7.240}$$

and

$$P = \int_{U_*}^{U^*} \left\{ \int_{U_*}^{u} 2h(u')\,\mathrm{d}u' \right\}^{-\frac{1}{2}} \mathrm{d}u. \tag{7.241}$$

We observe that the maximum concentration increases with the breadth of this dish to a limit $U_{\max}$ given by

$$\int_0^{u_e} h(u)\,\mathrm{d}u = \int_{u_e}^{U_{\max}} -h(u)\,\mathrm{d}u.$$

In case Δ_1 and Δ_2 are functions of u and v, eqn (7.236) could be satisfied by the solution of the ordinary differential equation

$$\frac{\mathrm{d}v}{\mathrm{d}u} = \frac{\Delta_2(u,v)}{\Delta_1(u,v)},$$

with $v = v_e$ when $u = u_e$. With this solution, say $v = V(u; u_e, v_e)$, we would take

$$h(u) = g(u) - f(u)V(u; u_e, v_e)$$

and proceed on similar lines. Such would be the case were Δ_2 proportional to v.

7.12. Some aspects of morphogenesis

In 1952 A. M. Turing suggested that some of the main phenomena of morphogenesis could be explained on the basis of the diffusion and reaction within the tissue of a system of chemical substances, called morphogens. The morphogens are initially distributed in a uniform manner, but if this uniform state became, at some stage of development, unstable it could lead to a stable non-uniform steady state from which the patterns of observed growth and form might then emerge. Turing's pioneer work was unfortunately cut off by his untimely death, but his ideas have been developed by Scriven and others in a penetrating analysis of both the stability and symmetry properties of the equations. We shall give some indication of the route which this work has followed though a comprehensive treatment would require a monograph in itself.

Turing considered a ring of cells, each of which was connected to its two neighbours, and Scriven and Gmitro (1966) showed that this structure, which Turing admitted was a little unrealistic, might be replaced by a one-dimensional system such as a rod or fibril. Similarly Turing's remarks on the sphere as a two-dimensional surface have been greatly extended to general surfaces and patterns of cells having certain symmetries. These structures are conceived as being the sites of a system of reactions between the morphogens, and to be accessible to a pool of morphogens by exchange through a membrane. It thus does not make sense to talk of a structure of more than two dimensions since the third physical dimension is needed for this pool. Equations for a general model of this kind can be set up by considering the factors that affect $c_i(n, t)$, the concentration of the ith morphogen A_i in the nth cell at time t. If the reactions between the S morphogens are

$$\sum_{i=1}^{S} a_{ji} A_i = 0 \quad j = 1, 2, \dots, R$$

and the rate of the jth reaction $r_j(c_1, \dots, c_s) = r_j(\mathbf{c})$, then the rate of formation A_i in a cell of volume V is

$$V \sum_{j=1}^{R} \alpha_{ji} r_j(\mathbf{c}).$$

If we suppose that the exchange of A_i is affected by all the concentration differences then it enters the cell from the pool at a rate

$$S_e \sum_{k=1}^{S} h_{ik}\{c_{kf} - c_k(n, t)\},$$

where S_e is the area for external exchange. It also exchanges with its neighbours, whose cell numbers will be denoted by p, at a rate

$$S_i \sum_p \sum_{k=1}^{S} h'_{ik}\{c_k(p, t) - c_k(n, t)\},$$

where S_i is the area for exchange with adjacent cells. Thus

$$\begin{aligned} V\frac{dc_i}{dt} = S_i \sum_p \sum_{k=1}^{S} h'_{ik}\,(c_k(p, t) - c_k(n, t)) + \\ + S_e \sum_{k=1}^{S} h_{ik}\{c_{kf} - c_k(n, t)\} + V \sum_{j=1}^{R} \alpha_{ji} r_j(\mathbf{c}). \end{aligned} \tag{7.242}$$

Let h' be a quantity characteristic of the exchange coefficients, c_f a typical concentration, and

$$\begin{aligned} u_i = c_i/c_f, \quad u_{if} = c_{if}/c_f, \quad \delta_{ik} = S_i h'_{ik}/S_e h', \quad \nu_{ik} = h_{ik}/h' \\ \tau = S_e h' t/V, \quad \psi_j^2 = V r_j(\mathbf{c}_f)/S_e c_f h', \quad R_j = r_j(\mathbf{c})/r_j(\mathbf{c}_f); \end{aligned} \tag{7.243}$$

then

$$\frac{du_i}{d\tau} = \sum_p \sum_k \delta_{ik}\{u_k(p, \tau) - u_k(n, \tau)\} + \sum_k \nu_{ik}(u_{kf} - u_k) + \sum_j \alpha_{ji}\psi_j^2 R_j(\mathbf{u}) \tag{7.244}$$

This is a lumped model appropriate to the discrete nature of the cell and it could clearly be analysed by the methods appropriate to ordinary differential equations. Turing (1952) considered the ring of cells for which the neighbours p of cell n would be those of index $(n-1)$ and $(n+1)$. Scriven and Othmer (1971) showed how to treat any lattice array of cells using group theory to simplify the equations. If the cells are small there may be merit in formulating a continuous analogue of these equations in which the difference operator over neighbouring cells would become the Laplacian and the cell parameters ψ_j would be Thiele moduli of the reactions ϕ_j. We would thus have

$$\frac{\partial u_i}{\partial \tau} = \sum_k \Delta_{ik}\nabla^2 u_k + \sum_k \nu_{ik}(u_{kf} - u_k) + \sum_j \alpha_{ji}\phi_j^2 R_j(\mathbf{u}), \tag{7.245}$$

a set of equations differing from those we have had before in the second group of terms, representing the transfer from the pool. A steady state of these equations will satisfy

$$0 = \sum_k \Delta_{ik}\nabla^2 u_{ke} + \sum_k \nu_{ik}(u_{kf} - u_{ke}) + \sum_j \alpha_{ji}\phi_j^2 R_j(\mathbf{u}_e), \tag{7.246}$$

and the small departures from steady state

$$u_i^*(\boldsymbol{\rho}, \tau) = u_i(\boldsymbol{\rho}, \tau) - u_{ie}(\boldsymbol{\rho}) \tag{7.247}$$

the linear equations

$$\frac{\partial u_i^*}{\partial \tau} = \sum_k \Delta_{ik} \nabla^2 u_k^* + \sum_k \kappa_{ik} u_k^*, \tag{7.248}$$

where

$$\kappa_{ik} = \nu_{ik} + \sum_j \alpha_{ji} \phi_j^2 \left(\frac{\partial R_j}{\partial u_k} \right)_{\mathbf{u}=\mathbf{u}_e}. \tag{7.249}$$

In the matrix form this can be written

$$\frac{\partial \mathbf{u}^*}{\partial \tau} = \Delta \nabla^2 \mathbf{u}^* + \mathbf{K} \mathbf{u}^*. \tag{7.250}$$

Because of the presence of the exchange terms ν_{ik}, some of which may indeed be representing active transport, the matrix $\mathbf{K}$ need have none of the definiteness properties that the monomolecular system of Section 5.2 had. The assumption of microscopic reversibility gives to the reaction matrix alone a certain symmetrizability, but unless peculiar circumstances obtain the $\mathbf{K}$ of eqn (7.250) will have no such properties nor will it be simultaneously diagonalizable with $\boldsymbol{\Delta}$.

By expanding each component of $\mathbf{u}^*$ in the eigenfunctions of the Laplace operator for the region being studied, i.e. setting

$$u_i^*(\boldsymbol{\rho}, \tau) = \sum_{n=1}^{\infty} a_{ni}(\tau) w_n(\boldsymbol{\rho}),$$

or

$$\mathbf{u}^*(\rho) = \sum \mathbf{a}_n(\tau) w_n(\boldsymbol{\rho}),$$

where

$$\nabla^2 w_n + \lambda_n^2 w_n = 0,$$

we obtain equations for the amplitude vector

$$\frac{\mathrm{d}\mathbf{a}_n}{\mathrm{d}\tau} = (\mathbf{K} - \lambda_n^2 \boldsymbol{\Delta}) \mathbf{a}_n. \tag{7.251}$$

If the region under consideration is unbounded as with a filament of infinite length or the whole plane λ can have any real value and a pattern of wavelength or characteristic size $2\pi/\lambda$ is obtained. If the region is finite there will be a discrete set of λ_n and a maximum characteristic pattern size as with the circular patch in the preceding section. The growth or decay of the amplitude function will depend on the eigenvalues of the matrix $\mathbf{K} - \lambda^2 \boldsymbol{\Delta}$. Turing gave examples with two and three morphogens to show the types of disturbance that could be obtained, and Scriven and Othmer have analysed these in detail (1969). With two species the eigenvalues are the roots of a quadratic and so can be obtained explicitly. For the cubic obtained when there are three species the analysis is more painful, but can again be done exhaustively. For more than three species a full parametric study is out of the question but Scriven and Othmer have shown how some progress can be made if the system can be decomposed into subsystems of two or three components that are only weakly coupled.

An important simple example of a bounded region arises when we consider the surface of the sphere for which

$$\nabla^2 = \frac{1}{\alpha^2 \sin^2\theta}\frac{\partial^2}{\partial\phi^2} + \frac{1}{\alpha^2 \sin\theta}\frac{\partial}{\partial\theta}\left(\sin\theta\frac{\partial}{\partial\theta}\right), \tag{7.252}$$

where α is the dimensionless radius of the sphere. The appropriate expansion of $\mathbf{u}^*$ is then

$$\mathbf{u}^* = \sum_{n=0}^{\infty}\sum_{m=-n}^{n} \mathbf{a}_{mn} P_n^m(\cos\theta)\, e^{im\phi} \tag{7.253}$$

where P_n^m is the associated Legendre function of the first kind. Then

$$\frac{d\mathbf{a}_{mn}}{d\tau} = \left\{\mathbf{K} - \frac{n(n+1)}{\alpha^2}\mathbf{\Delta}\right\}\mathbf{a}_{mn}. \tag{7.254}$$

Since n can be zero the steady state solution will only be stable if $\mathbf{K}$ is negative definite. Suppose that $\mathbf{K}$ has an eigenvalue of positive real part, then a uniform perturbation will grow, but the system may still be stable to higher modes of perturbation ($n \geqslant 1$) provided the sphere is sufficiently small. However, the system will become unstable to all higher modes of perturbation when the radius becomes sufficiently large. For suppose that there is a critical value λ_c such that $(\mathbf{K} - \lambda^2\mathbf{\Delta})$ is negative definite for $\lambda > \lambda_c$ but not for $\lambda < \lambda_c$. Then as soon as α exceeds $\{n(n+1)\}^{\frac{1}{2}}/\lambda_c$ the system will become unstable to the nth mode.

A more interesting case arises when $\mathbf{K}$ is negative definite, so that the system is stable to uniform perturbations, and $\mathbf{\Delta}$ is positive definite but there is a range of values of λ, say (λ_b, λ^b), for which $(\mathbf{K} - \lambda^2\mathbf{\Delta})$ has a positive eigenvalue. Scriven and Othmer (1969) have given examples of this, their case (a), with two species: the movement of the two eigenvalues is shown in the top part of Fig. 7.25. In this situation those modes alone will be unstable for which

$$(\lambda_b\alpha)^2 < n(n+1) < (\lambda^b\alpha)^2.$$

Thus it might well be that as the radius of the sphere increases a patterned perturbation grows on its surface. This is shown in the middle part of the

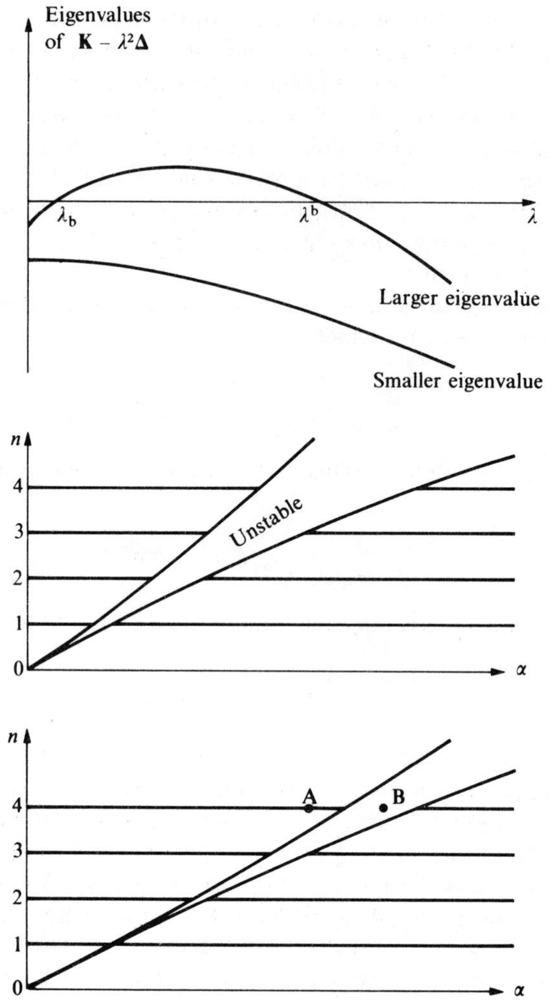

FIG. 7.25. Illustrating the origin of an unstable pattern of the nth kind during growth.

figure where mode n is stable for all values of α except those in the interval

$$\frac{\sqrt{\{n(n+1)\}}}{\lambda^b} < \alpha < \frac{\sqrt{\{n(n+1)\}}}{\lambda_b}.$$

If λ_b and λ^b are close together, these size intervals may be quite separate as shown in the bottom part of Fig. 7.25. Suppose that as the sphere grows, the matrices $\mathbf{K}$ and $\mathbf{\Delta}$ fall into this class of behaviour when the size reaches a value such as is indicated by the point A. It is here stable to all modes of perturbation but as it grows still more (point B) it reaches a state which is unstable to any mode of perturbation with $n = 4$. If there is some stable solution of the non-linear equations having this symmetry, a chance perturbation might then give rise to a patterned distribution of the morphogens and trigger an entirely new form of development away from spherical form. Clearly with an increasing number of morphogens there arise even more possibilities for behaviour of this type.

These considerations also lead us to recognize the importance of non-uniform steady state solutions of the diffusion and reaction equations which have specific symmetry properties. It is to the study of these that Gmitro's work brings the full power of group theoretic methods (Gmitro (1969)). He uses symmetry adapted harmonics to provide the most economical set of equations for the amplitude functions, in building up an approximation to the solution of the equation which has the symmetry of a given group. The method is analogous to Eckhaus' finite amplitude technique, and the application of group theory ensures that only those solutions with the desired symmetry appear and that they do so as economically as possible. As a particular example he has shown that the nonlinear equation in just one concentration

$$\frac{\partial u}{\partial \tau} = \nabla^2 u + \phi^2 R(u), \tag{7.255}$$

with

$$R(u) = \frac{u(1+\alpha u)}{1+\alpha}, \tag{7.256}$$

can have solutions on the surface of a sphere which tend to a stable, non-uniform steady state with octahedral symmetry for suitable values of the parameters ϕ and α. Figure 7.26 shows the contours of one of the symmetry adapted harmonics that Gmitro uses in obtaining this solution.

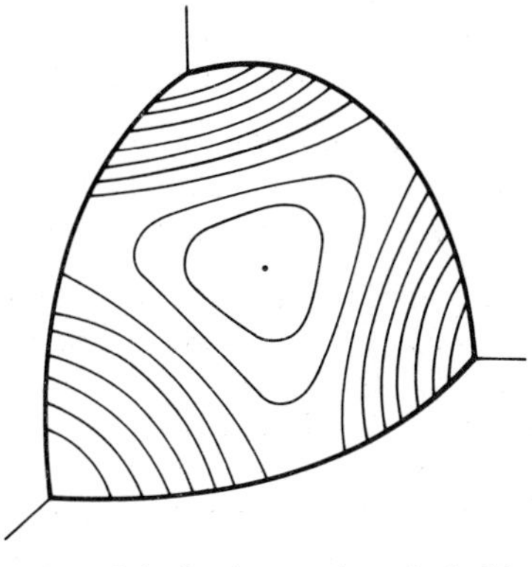

FIG. 7.26. Lines of constant value of the fundamental octahedral harmonic on the first octant of a sphere.

This work is of great depth and importance for the understanding of pattern formation, but with this very bare introduction we must take leave of the topic and close this survey of some of the aspects of stability and instability of the steady state.

Additional Bibliographical Comments

7.1. Chemical reactors are of course instances of dynamical systems for which there is a large body of theory, see, for example, Bahtia and Szego (1970). An elementary exposition of this idea is given by Aris (1971*b*, 1972*b*). Gall (1972) has applied some ideas from non-linear functional analysis to reactor stability.

Perlmutter's book on the stability of chemical reactors (1972) is a very good general survey: Denn's more recent book (1974) is more comprehensive and is particularly valuable in relating different physical problems to the important mathematical techniques. Mel (1964) has considered a fluid dynamical stability problem in which a continuous steady state enzymic reaction is controlled by gravity and the diffusion of inerts.

7.2. Justification of linearization is discussed by Krasnosel'skii (1964). Villadsen and Michelsen (1972) report a case in which the middle steady state of five solutions of the Robin problem is stable.

7.3. Some basic references on maximum principle or comparison theorems are: Westphal (1949); Prodi (1951); Szarski (1955, 1965); Mlak (1957); Narsimhan (1954); McNabb (1961); Friedman (1966); Sattinger (1968, 1972, 1974); Kastenberg (1967, 1970). Some cognate problems having to do with nuclear reactors are considered in: Kastenberg and Chambré (1968); Kastenberg (1969).

7.4. Other papers using the Liapounov method are: Fjeld (1967); Lapidus and Berger (1968); Nishimura and Matsubara (1969); Murphy and Crandall (1970). Perlmutter and McGowin (1971) combine it with a collocation technique. The basic theorems are proved in Lakshmikanthan (1964) and some later results are given by Auchmuty (1973). There is a considerable literature on the application of the method for ordinary differential equations to chemical reactors. An excellent guide to the generation of Liapounov functions is given by Lapidus and Gurel (1969).

7.5 The results of Hlaváček and Marek (1968*c*) concerning limit cycles will be discussed in Section 8.2. Oscillations about a unique steady state were also found by Schmitz and Lindberg (1971) for reaction on a surface near a stagnation point.

7.6. The stability of the reaction on the surface of an impermeable catalytic sphere has been discussed by Petersen and Friedly (1964) using a linearization technique. Petersen, Friedly, and DeVogelaere (1964) gave exact solutions both for uniform conditions at infinity and for a distant gradient. The dynamics of a surface reaction near a stagnation point have been considered by Schmitz and Winegardner (1967) and the multiplicity of states by Schmitz and Lindberg (1970). Schmitz has also studied the diffusion flame, whose basic equations differ only by a convective term from those of catalytic reaction in a slab; see Schmitz and Grosboll (1965), Schmitz and Kirkby (1966) and Schmitz (1967).

7.7. Hlaváček and his colleagues have done much to develop the analogy of the stirred tank: see Hlaváček and Marek (1968*a*), Hlaváček, Marek and Kubíček (1969*a*, *b*, 1970), Hlaváček (1970*a*). A collocation method is used in Hlaváček and Kubíček (1971*a*) and various methods are compared in their papers (1971*b*, *c*). Finlayson and Ferguson (1970) is a careful study of collocation much of which is incorporated in the section of Finlayson (1972) referred to above. Amundson and Raymond (1965) and Amundson and Kuo (1967*a*, *b*) consider models with resistance to transfer of either heat or mass lumped at the surface. Lapidus, Lee and Padmanabhan consider the isothermal slab with external resistances.

7.10. The equations for the surface reaction with external diffusion have much in common with those for tunnel diode circuits: see Landauer (1962, 1971); Wang (1973).

7.11. Further references may be found in the papers of Segel and Keller (1971*a*, *b*), Segel (1971), Segel and Stoeckly (1972).

7.12. Scriven's work with Othmer and Gmitro is to be found in Scriven and Othmer (1969, 1971, 1973), Scriven and Gmitro (1966), Gmitro (1969), Othmer (1969). Turing's scheme of six components is considered in another context by Prigogine and Nicolis (1967). Rosen (1970) gives an excellent review of the bearing of the theory of dynamical systems on biology. See also Wardlaw (1953, 1955), Thom (1968). Crick's letter in *Nature* (1970), following a paper of Wolpert (1969), has been very influential in drawing attention to the importance of diffusion in embryogenesis. A discussion of bifurcation in the presence of a symmetric group is given by Ruelle (1973).

The question of the variation of stability with variations of a small Lewis number is allied to the effect of the small parameter in singularly perturbed systems; this has been studied in a chemical context by Amundson, Schneider, and Aris (1973) and as so-called parasitics in circuit theory by Shensa (1971), Shensa and Desoer (1970), Chua and Alexander (1971); see also Zien (1973).

NOMENCLATURE

(See also the General Nomenclature, p. xii)

A	region of asymptotic stability (Section 7.10)
A_c, A^c	critical amplitudes (Section 7.8)
$\mathbf{A}, \mathbf{B}$	matrices defined by eqn (7.91)
A_j, B_{ij}	collocation constants (Section 7.7)
A_n, a_{ni}	amplitudes
$\mathbf{a}_{mn}, \mathbf{a}_n$	vectors of amplitudes (Section 7.12)
a, b	constants in model problem, eqn (7.173)
$\mathbf{C}, \mathbf{D}$	matrices defined by eqn (7.128)
F, f_+, f_-	non-linear equation and boundary functions in the Peterson–Maple theorem (Section 7.10)
$\mathcal{F}$	function defining deviation of reaction rate, eqn (7.50)
f, g	production and degradation rates of acrasin (Section 7.11)
G	non-linear equation in the Westphal–Prodi theorem, eqn (7.55)

g, h	$\phi^2(\beta R_v - R_u)$, $\phi^2 R_u$ in Section 7.4
h_{ik}, h'_{ik}	exchange coefficients in cell model of morphogenesis
$\mathbf{J}, \mathbf{K}$	Jacobian matrix, matrix of its eigenvectors, eqn (7.155)
L	capacity parameter for catalytic wire (Section 7.9)
M_j	deviation from the mean defined by eqn (7.166)
M_{ij}, N_{ij}	matrix elements derived from collocation matrices eqn (7.152)
m, n	indices of surface harmonic modes (Section 7.12)
$\mathbf{N}$	matrix defined by eqn (7.128)
$R^*(u^*, v^*, \boldsymbol{\rho})$	$R(u_e(\boldsymbol{\rho}) + u^*, v_e(\boldsymbol{\rho}) + v^*) - R(u_e, v_e)$
$\mathbf{R}, \mathbf{S}$	matrices defined in eqn (7.159)
S_e, S_i	external and internal interchange areas in cell model (Section 7.12)
s	$\{s_1^2 + s_2^2\}^{\frac{1}{2}}$, wave number (Section 7.11)
U^*, U_*	bounds defined by eqn (7.240)
u	concentration of acrasin (Section 7.11)
$u_i(n, \tau)$	concentration of ith morphogen in cell n (Section 7.12)
$V(\tau)$	Liapounov functional
v	concentration of cells (Section 7.11)
$\dot{v}_e$	$\partial v_e/\partial \phi$ (Section 7.2)
$v_1, v_{1+\beta}$	solutions with initial conditions $v_0 = 1$ and $1+\beta$, respectively (Section 7.3)
w	comparison variable
$\mathbf{w}$	vector of concentration and temperature deviations (Section 7.6)
w_n	eigenfunctions of the Laplacian in Ω
A_c	area of circular patch of critical radius P_c (Section 7.11)
α	dimensionless radius (Section 7.12)
Γ_*, Γ^*	types of region in the (u_+, u_-)-plane (Section 7.10)
Δ_1, Δ_2	chemotactic and ordinary diffusion coefficients (Section 7.11)
$\boldsymbol{\Delta}$	matrix of multicomponent diffusion coefficients (Section 7.12)
δ_{ik}	internal exchange coefficients (Section 7.12)
Z, ζ	total length and length variable for catalytic wire (Section 7.9)
$\mathbf{K}$	matrix of reaction-exchange coefficients (Section 7.12)
κ_{ik}	elements defined by eqn (7.249)
K_i, κ_i	overall and internal transfer coefficients eqn (7.164)
λ_b, λ^b	bounds on the region of unstable eigenvalues (Section 7.12)
λ_n^2	eigenvalues of Laplacian in Ω, cf. eqn (7.64)
μ	eigenvalue
μ', ν'	modified Biot numbers for catalytic wire (Section 7.9)
ν_{ik}	external exchange coefficients (Section 7.12)

P	breadth of dish in chemotactic clumping solution
P_c	critical radius for instability (Section 7.11)
ψ^2	analogue of Thiele modulus for stirred tank
χ^2	adjustable parameter in Liapounov functional
$\Omega_{0,T}$	$\{\boldsymbol{\rho}, \tau; \boldsymbol{\rho} \in \Omega, 0 < \tau < T\}$ region defined in Section 7.3

SUFFIXES AND AFFIXES

*	deviation from steady state
e	steady state or equilibrium solution
+, −	pertaining to boundaries at $\rho = \pm 1$ (Section 7.10)
$_b$, ^{b}	lower and upper bounds of a region of non-uniqueness or instability
$\langle . \rangle$	mean value over Ω.

8

SOME FEATURES OF THE TRANSIENT BEHAVIOUR OF DIFFUSING AND REACTING SYSTEMS

IN this chapter we shall touch on some of the problems that arise in describing the transients for diffusing and reacting systems. We have of course encountered time-dependent solutions in discussing stability, but there remain some rather interesting problems that it will be useful to bring together in one place. Nearly all are the subject of continuing research and it is impossible to give more than a general introduction to their peculiar features. They include the use of deliberate unsteady state conditions by varying the boundary conditions periodically, poisoning, and other forms of decay, extremely rapid variations of temperature for which the validity of the usual equations is in doubt and some problems suggested by biological considerations.

One interesting problem, which we shall not describe in detail, is the build-up of pressure within a porous catalyst particle which is wet by a liquid reactant. This has been thoroughly analysed by Sangiovanni and Kesten (1971), and they have compared their results with some observations on a hydrazine reactor. When a liquid reactant comes into contact with a porous catalyst pellet, it is drawn into the pores by capillary action and prevents the escape of gaseous products until sufficient pressure builds up and expels the liquid. This can cause delayed ignition and a subsequent sharp increase of pressure when the reactor is started up.

8.1. Linear problems

A general linear problem in a single dependent variable is given by

$$\frac{\partial u}{\partial \tau} = \nabla^2 u - \phi^2 u \tag{8.1}$$

in Ω subject to

$$\frac{1}{\nu}\frac{\partial u}{\partial \nu} + u = 1 \tag{8.2}$$

on Ω and the initial condition

$$u(\boldsymbol{\rho}, 0) = u_0(\boldsymbol{\rho}). \tag{8.3}$$

The steady state solution

$$\lim_{\tau \to \infty} u(\boldsymbol{\rho}, \tau) = u_e(\boldsymbol{\rho}) \tag{8.4}$$

satisfies the first two equations with the time derivative set equal to zero. Then the solution of the full transient problem is

$$u(\boldsymbol{\rho}, \tau) = u_e(\boldsymbol{\rho}) + \sum_{n=1}^{\infty} a_n w_n(\boldsymbol{\rho}) \exp\{-(\lambda_n^2 + \phi^2)\tau\} \tag{8.5}$$

where λ_n^2 and $w_n(\boldsymbol{\rho})$ are the eigenvalues and normalized functions of the problem

$$\nabla^2 w + \lambda^2 w = 0 \tag{8.6}$$

in Ω,

$$\frac{1}{v}\frac{\partial w}{\partial v} + w = 0 \tag{8.7}$$

on $\partial\Omega$, and

$$a_n = \frac{1}{\upsilon} \iiint_{\Omega} \{u_0(\boldsymbol{\rho}) - u_e(\boldsymbol{\rho})\} w_n(\boldsymbol{\rho}) \, \mathrm{d}\Upsilon. \tag{8.8}$$

The effectiveness factor at time τ is related to the steady state value by

$$\eta(\tau) = \eta + \sum_{n=1}^{\infty} a_n \langle w_n \rangle \exp\{-(\lambda_n^2 + \phi^2)\tau\} \tag{8.9}$$

where $\langle \, . \, \rangle$ denotes the average value over Ω.

Danckwerts (1951) used an alternative form which is useful when $u_0(\boldsymbol{\rho}) = 0$. If $w(\boldsymbol{\rho}, \tau)$ is the solution of eqns (8.1) and (8.2) when $\phi = 0$ and $w(\boldsymbol{\rho}, 0) = 0$, then

$$\begin{aligned} u(\boldsymbol{\rho}, \tau) &= w(\boldsymbol{\rho}, \tau) \exp(-\phi^2 \tau) + \phi^2 \int_0^{\tau} w(\boldsymbol{\rho}, \sigma) \exp(-\phi^2 \sigma) \, \mathrm{d}\sigma \\ &= \int_0^{\tau} \left(\frac{\partial w}{\partial \sigma}\right) \exp(-\phi^2 \sigma) \, \mathrm{d}\sigma. \end{aligned} \tag{8.10}$$

This may be confirmed by direct substitution as is shown by Crank (1956) who also gives a number of different instances of its application.

The Laplace transform may also be used to reduce the equation to an elliptic one. Thus if

$$\bar{u}(\boldsymbol{\rho}, s) = \int_0^{\infty} \mathrm{e}^{-s\tau} u(\boldsymbol{\rho}, \tau) \, \mathrm{d}\tau \tag{8.11}$$

and $u_0(\boldsymbol{\rho}) = 0$, we would have

$$\nabla^2 \bar{u} - (\phi^2 + s)\bar{u} = 0 \tag{8.12}$$

in Ω, with

$$\frac{1}{\nu}\frac{\partial \bar{u}}{\partial \nu}+\bar{u} = \frac{1}{s} \tag{8.13}$$

on $\partial\Omega$. If $u_e(\boldsymbol{\rho}\,;\phi^2, \nu)$ denotes the solution of the steady state problem defined by eqns (8.1) and (8.2) then

$$\bar{u}(\boldsymbol{\rho}, s) = u_e(\boldsymbol{\rho}\,;\phi^2+s, \nu)/s. \tag{8.14}$$

By the final value theorem the Laplace transform

$$\lim_{\tau\to\infty} u(\boldsymbol{\rho}, \tau) = \lim_{s\to 0} s\bar{u}(\boldsymbol{\rho}, s) = u_e(\boldsymbol{\rho}\,;\phi^2, \nu),$$

which is as it should be. More usefully, if $\bar{\eta}(s)$ is the Laplace transform of $\eta(\tau)$ and $\eta(\phi^2, \nu)$ the steady state value then

$$s\bar{\eta}(s) = \eta(\phi^2+s, \nu) \tag{8.15}$$

and

$$\eta(\tau) = \frac{1}{2\pi i}\int_{\gamma-i\infty}^{\gamma+i\infty} \eta(\phi^2+s, \nu)\frac{e^{s\tau}}{s}\,ds, \tag{8.16}$$

where γ is in the half-plane of analyticity of $\eta(\phi^2+s, \nu)$. The Laplace transform in eqn (8.15) may also be interpreted as

$$\eta(\tau) = \int_0^{\tau} \exp(-\phi^2\sigma)H_\nu(\sigma)\,d\sigma, \tag{8.17}$$

where

$$L[H_\nu(\tau)] = \eta(s, \nu). \tag{8.18}$$

If the concentration c_f far from the surface $\partial\Omega$ is not constant, then some characteristic value must be chosen to render the variables dimensionless and the boundary condition (8.12) written in the form

$$\frac{1}{\nu}\frac{\partial u}{\partial \nu}+u = u_f(\tau). \tag{8.19}$$

The Laplace transform treatment, still with $u_0 = 0$, leads again to eqn (8.12) with the boundary condition

$$\frac{1}{\nu}\frac{\partial \bar{u}}{\partial \nu}+\bar{u} = \bar{u}_f(s) \tag{8.20}$$

and

$$\bar{u}(\boldsymbol{\rho}, s) = \bar{u}_f(s)u_e(\boldsymbol{\rho}, \phi^2+s, \nu). \tag{8.21}$$

If $u_1(\mathbf{\rho}, \tau)$ is the solution when $u_f(\tau) \equiv 1$, then the convolution theorem gives

$$u(\mathbf{\rho}, \tau) = \int_0^\tau u_f(\sigma) u_1(\mathbf{\rho}, \tau - \sigma)\, d\sigma. \tag{8.22}$$

The connections with the case of no reaction allow the many known solutions of conduction and diffusion problems to be adapted to the case of first-order reaction. The splendid treatises of Carslaw and Jaeger (1959) and Crank (1956) are of course the great source of reference material on this score.

Crank (1956) gives the solution to several other problems involving reaction. Of particular importance is the first order reversible reaction in which the product is immobilized in the porous material. Such would be the situation with the dyeing of a fabric or impregnation of a catalyst. If u is the concentration of reactant (mobile dye) and v that of the product (immobilized dye), then

$$\frac{\partial u}{\partial \tau} = \nabla^2 u - \phi^2 (u - \lambda v), \tag{8.23}$$

$$\frac{\partial v}{\partial \tau} = \phi^2 (u - \lambda v) \tag{8.24}$$

in Ω. If the source of reactant (or dye) is a well-stirred container of m times the volume of Ω, then the concentration $u_f(\tau)$ will vary, and

$$\frac{1}{v} \frac{\partial u}{\partial v} + u = u_f(\tau) \tag{8.25}$$

on $\partial\Omega$ while $u_f(\tau)$ itself is governed by

$$mv \frac{du_f}{d\tau} = \iint_{\partial\Omega} \left(\frac{\partial u}{\partial v}\right) d\Sigma. \tag{8.26}$$

This last equation may be translated into an overall balance by adding the sum of eqns (8.23) and (8.24) integrated over Ω; then

$$mu_f + \langle u \rangle + \langle v \rangle = m \tag{8.27}$$

merely expresses the conservation of a substance that was initially to be found at unit concentration only in the surrounding chamber. The final steady state is uniform with

$$u = u_f = \lambda v = \lambda m / (1 + \lambda + \lambda m). \tag{8.28}$$

Taking the Laplace transform of eqns (8.23–6) and assuming $u(\mathbf{\rho}, 0) =$

$= v(\boldsymbol{\rho}, 0) = 0$ we have

$$\nabla^2\bar{u} - (\phi^2 + s)\bar{u} + \phi^2\lambda\bar{v} = 0, \tag{8.29}$$

$$\phi^2\bar{u} - (\lambda\phi^2 + s)\bar{v} = 0 \quad \text{in } \Omega, \tag{8.30}$$

$$\frac{1}{v}\frac{\partial\bar{u}}{\partial v} + \bar{u} = \bar{u}_{\mathrm{f}} \quad \text{on } \partial\Omega, \tag{8.31}$$

and

$$mv(s\bar{u}_{\mathrm{f}} - 1) = -\iint\limits_{\partial\Omega}\left(\frac{\partial\bar{u}}{\partial v}\right)\mathrm{d}\Sigma = -v\sigma(\bar{u}_{\mathrm{f}} - \langle\bar{u}\rangle_{\mathrm{s}}), \tag{8.32}$$

where $\langle\bar{u}\rangle_{\mathrm{s}}$ denotes the average of the surface value of $\bar{u}$. In the case of symmetrical shapes for which $\langle\bar{u}\rangle_{\mathrm{s}} = \bar{u}_{\mathrm{s}}$ we can eliminate $\bar{v}$ and $\bar{u}_{\mathrm{f}}$ from the equations. For on setting $\mu = (v\sigma/mv)$ eqn (8.32) gives

$$\bar{u}_{\mathrm{f}} = \frac{1 + \mu\bar{u}}{s + \mu}$$

and on substitution into eqn (8.31) we have

$$\frac{1}{v}\frac{\partial\bar{u}}{\partial v} + \bar{u}\left(\frac{s}{s+\mu}\right) = \frac{1}{s+\mu}. \tag{8.33}$$

On the other hand, substituting for $\bar{v}$ from eqn (8.30) into eqn (8.29) gives

$$\nabla^2\bar{u} - \left(\frac{\phi^2 s}{\lambda\phi^2 + s} + s\right)\bar{u} = 0. \tag{8.34}$$

It follows that

$$s\bar{u}(\rho, s) = u_{\mathrm{e}}\left(\rho, \frac{\phi^2 s}{\lambda\phi^2 + s} + s, \frac{vs}{s+\mu}\right) \tag{8.35}$$

with the corresponding formulae for $\bar{v}$ and $\bar{u}_{\mathrm{f}}$. The total amount of solute taken up into Ω at time τ, is of interest. Let

$$W(\tau) = \frac{1 + \lambda + \lambda m}{(1+\lambda)m}\iiint\limits_{\Omega}(u + v)\,\mathrm{d}\Upsilon \tag{8.36}$$

be the fraction of solute so taken up at time τ, then

$$\overline{W}(s) = \frac{1 + \lambda + \lambda m}{(1+\lambda)ms}\left(1 + \frac{\phi^2}{\lambda\phi^2 + s}\right)\eta\left(\frac{\phi^2 s}{\lambda\phi^2 + s} + s, \frac{vs}{s+\mu}\right), \tag{8.37}$$

where $\eta(\phi^2, v)$ is the effectiveness factor for the steady state problem. These formulae, whilst compact, have to be inverted to give the required solution in the time domain. They do however demonstrate a valuable connection between the solutions of diffusion problems with and without reaction.

8.2. Limit cycles and other transient phenomena

In discussing the effect of the Lewis number on the stability of the steady state we had occasion to look at some of the transient solutions that have been calculated for the slab and sphere. Apart from the work of Luss and Lee there reported (Section 7.5), the most extensive studies have been those of Hlaváček, Marek and Kubíček (1969*a*, *b*) They made use of the analogy with the stirred tank, first given by Frank-Kamenetskii, extensively developing it in their work. This, as we have seen, is one of a class of approximations which neglect all but the first term. We shall see in the next section that the analogy can also be justified by a perturbation method suitable for the case of a very large diffusion coefficient. The analogy also suggests a useful way of representing the transient behaviour as a trajectory in a phase plane of average concentration and temperature. A comparison between the phase planes of the distributed and lumped systems shows that they have the same qualitative features and trends, but that the actual numbers given by the approximation bear an unknown relation to those of the exact model.

Restricting attention to the single reaction we have (cf. Section 6.4.2) the distributed transient equations

$$\frac{\partial u}{\partial \tau} = \nabla^2 u - \phi^2 R(u, v), \tag{8.38}$$

$$\mathscr{L}\frac{\partial v}{\partial \tau} = \nabla^2 v + \beta\phi^2 R(u, v) \tag{8.39}$$

in Ω, with

$$\frac{1}{\nu}\frac{\partial u}{\partial \nu} + u = 1, \qquad \frac{1}{\mu}\frac{\partial v}{\partial \nu} + v = 1 \tag{8.40}$$

on $\partial\Omega$. The analogous stirred tank model is

$$\frac{\mathrm{d}u}{\mathrm{d}\tau'} = 1 - \tilde{u} - \psi^2 R(\tilde{u}, \tilde{v}), \tag{8.41}$$

$$L\frac{\mathrm{d}\tilde{v}}{\mathrm{d}\tau'} = 1 - \tilde{v} + \beta'\psi^2 R(\tilde{u}, \tilde{v}), \tag{8.42}$$

where

$$\tau' = \lambda_1^2(\nu)\tau, \qquad \psi = \phi/\lambda_1(\nu), \; L/\mathscr{L} = \beta'/\beta = \lambda_1^2(\nu)/\lambda_1^2(\mu), \tag{8.43}$$

and $\lambda_1^2(\kappa)$ is the first eigenvalue of

$$\nabla^2 W + \lambda^2 W = 0 \quad \text{in } \Omega$$

$$\frac{\partial W}{\partial \nu} + \kappa W = 0 \quad \text{on } \partial\Omega. \tag{8.44}$$

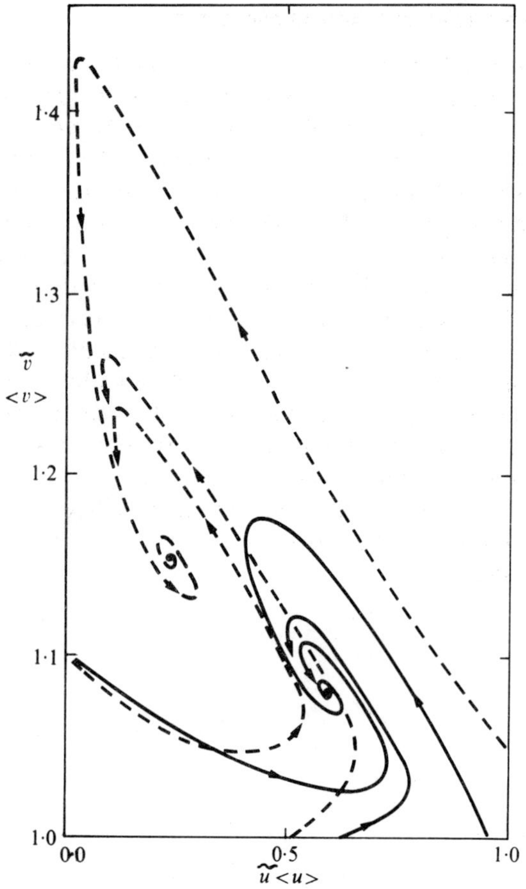

FIG. 8.1. Comparison of trajectories of the stirred tank (broken line) and catalyst particle (solid line) for $p = 1$, $q = 2$, $\beta = 0{\cdot}2$, $\gamma = 20$, $\phi = 1{\cdot}5$, $\mathscr{L} = 0{\cdot}4$, $\mu = \nu = \infty$. (After Hlaváček, Kubíček and Marek (1969*b*).)

The average concentrations and temperature which Hlaváček compares with $\tilde{u}$ and $\tilde{v}$ are simply the volume averages

$$\langle u \rangle = \frac{1}{\upsilon} \iiint u \, \mathrm{d}\Upsilon \quad \text{and} \quad \langle v \rangle = \frac{1}{\upsilon} \iiint v \, \mathrm{d}\Upsilon. \tag{8.45}$$

A weighted average using the first eigenfunction as the weighting function would serve to bring the curves rather closer together but the points are well made with simpler average.

Figure 8.1 shows a comparison of trajectories for a first-order, irreversible reaction in a sphere. The parameters in the equation and the familiar expression

$$R(u, v) = u^p \exp\{\gamma(v-1)/v\}$$

have the values

$$p = 1, \quad q = 2, \quad \beta = 0{\cdot}2, \quad \gamma = 20, \quad \phi = 1{\cdot}5, \quad \mu = \nu = \infty, \quad \mathscr{L} = 0{\cdot}4,$$

and the two eigenvalues are both equal, $\lambda_1 = \pi$. It is the case of a single steady state for which the Lewis number is still sufficiently large for stability. However the approach is an oscillatory one corresponding to a focus in the phase plane. For smaller values of the Lewis number the unique steady state becomes unstable and a limit cycle surrounds it. It is represented in the phase plane as a closed curve, and Fig. 8.2 shows how this limit cycle grows with

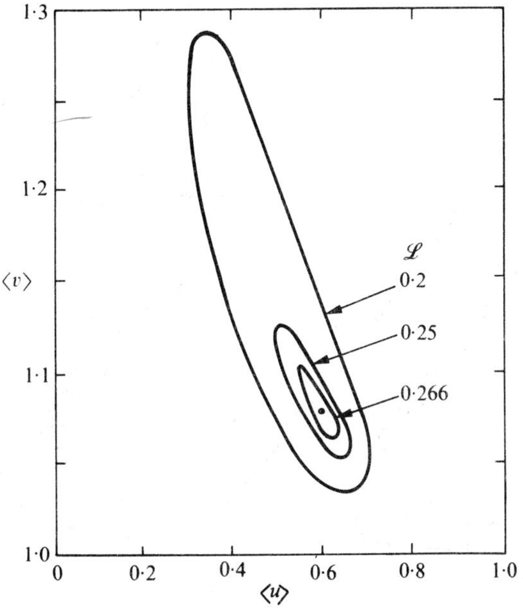

FIG. 8.2. Growth of the limit cycle with decreasing $\mathscr{L}$, the other parameters being the same as in Fig. 8.1. (After Hlaváček, Kubíček, and Marek, (1969*b*).)

decreasing $\mathscr{L}$. These curves are not traversed at a uniform rate, for the part at low temperatures is very slow compared with that at high temperatures. In particular we notice that these rapid excursions may be made to temperatures in excess of the Prater temperature. This is even more marked in the comparison shown in Fig. 8.3. The geometry for this calculation was that of

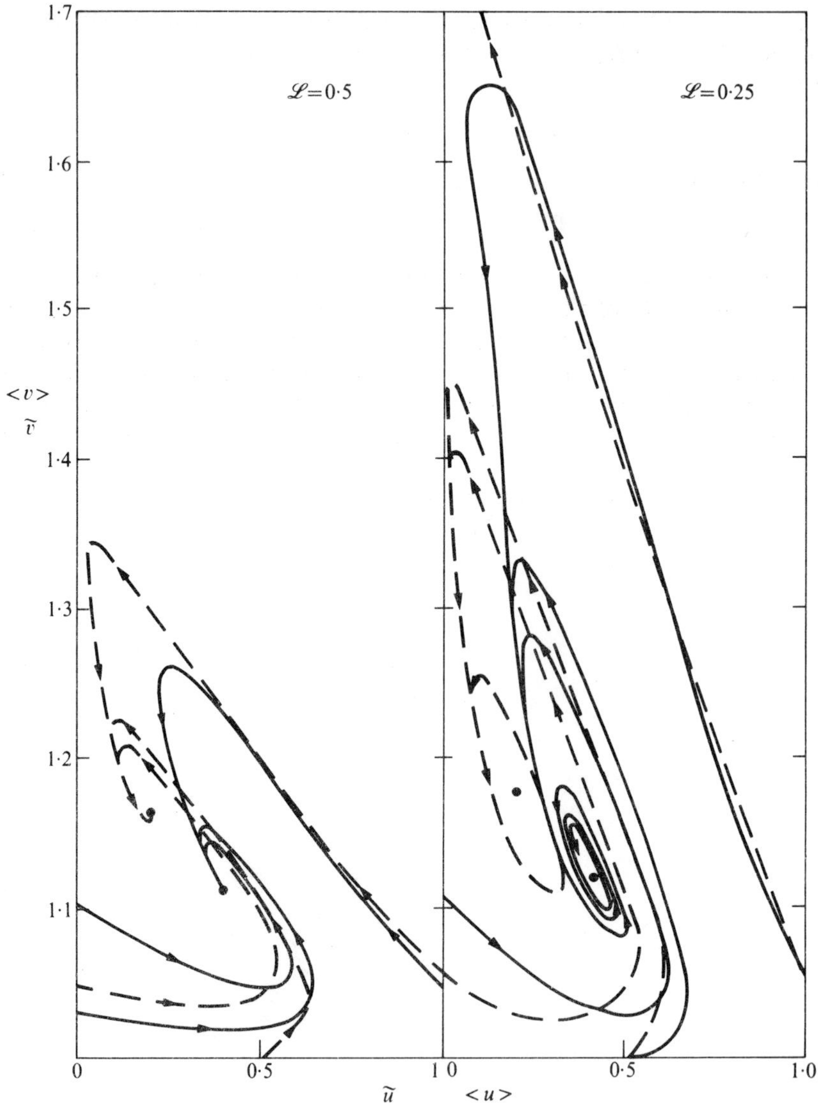

FIG. 8.3. Comparison of trajectories of the stirred tank (broken line) and catalyst particle (solid line) for $p = 1$, $q = 0$, $\beta = 0{\cdot}2$, $\gamma = 20$, $\phi = 0{\cdot}8$, $\mu = \nu = \infty$, and $\mathscr{L} = 0{\cdot}5$ and $0{\cdot}25$. (After Hlaváček, Kubíček, and Marek (1969*b*).)

the slab; the difference between the two parts of the figure being that of the Lewis number. As before the equivalent stirred tank exaggerates the temperature, but the trends are the same. However, we see that for the exact curves the temperature rises to more than three times the Prater temperature.

The critical value of L for the instability of a steady state $(\tilde{u}_e, \tilde{v}_e)$ of eqns (8.41) and (8.42) is

$$L_c = \frac{\beta'\psi^2 R_v(\tilde{u}_e, \tilde{v}_e) - 1}{1 + \psi^2 R_u(\tilde{u}_e, \tilde{v}_e)}, \tag{8.46}$$

and the steady state will be unstable for all $L < L_c$. In translating this into an equivalent condition on the Lewis number

$$\mathscr{L}_c = \frac{\beta\phi^2 R_v - \lambda_1^2(\mu)}{\lambda_1^2(\nu) + \phi^2 R_u}; \tag{8.47}$$

it is of course difficult to know what values to substitute for the arguments of R_u and R_v. Hlaváček was able to show tolerably good agreement between eqn (8.47) and a more accurately determined value when the average values $\langle u \rangle$ and $\langle v \rangle$ were used as arguments in the partial derivatives. As is always the case with this type of approximation the agreement is unpredictable and each case must be investigated on its own merits.

The phase plane is also instructuve when there are multiple solutions. Figure 8.4 shows three phase portraits in which only the minimal solution A is stable. For $\mathscr{L} = 0{\cdot}4$ a perturbation about the maximal solution C grows rapidly and the state approaches A in short order. For $\mathscr{L} = 0{\cdot}444$ the state makes three oscillations about C before breaking away to the stable steady

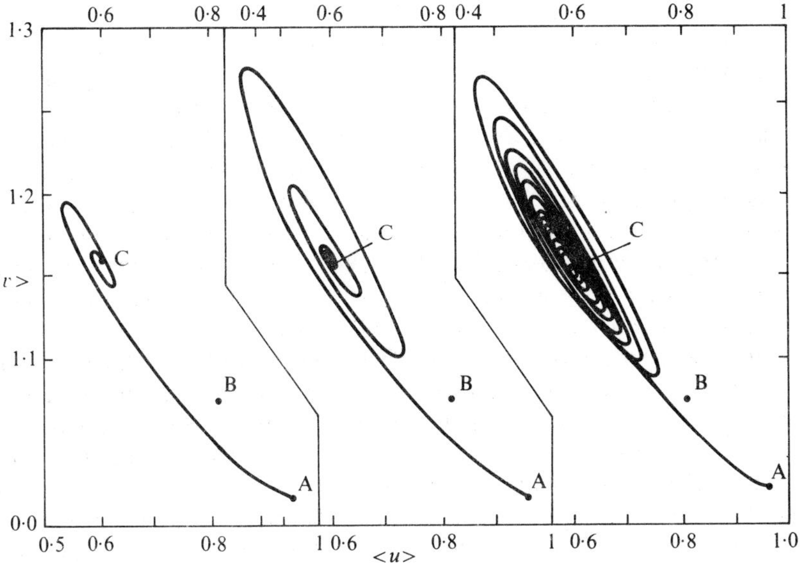

FIG. 8.4. Phase planes for a spherical catalyst particle with three steady states: $p = 1$, $q = 2$, $\beta = 0{\cdot}4$, $\gamma = 20$, $\phi = 0{\cdot}65$, $\mathscr{L} = 0{\cdot}4, 0{\cdot}444, 0{\cdot}472$. (After Hlaváček, Kubíček, and Marek (1969*b*).)

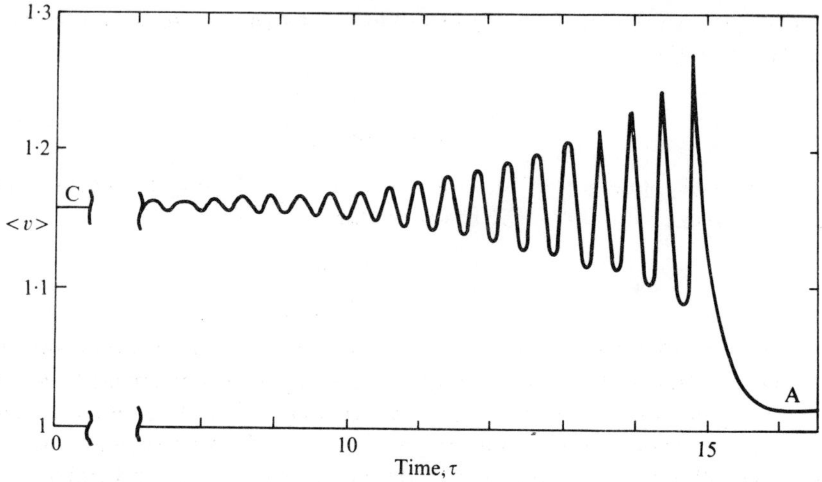

FIG. 8.5. The average temperature as a function of time for the trajectory shown in Fig. 8.4, $\mathscr{L} = 0{\cdot}472$. (After Hlaváček, Kubíček, and Marek (1969*b*).)

state. For $\mathscr{L} = 0{\cdot}472$ the maximal solution is only just unstable, and the system makes many oscillations before breaking away. This is shown graphically in Fig. 8.5, where $\langle v \rangle$ is seen as a function of time. Hlaváček and Marek (1968*c*) give many other transient curves, as do Luss and Lee (1970) in the context previously discussed (Section 7.5), but these suffice to show the general character of the behaviour.

Oscillations have been observed by Wicke and his colleagues both in isothermal and non-isothermal catalyst particles (Wicke and Fieguth 1971; Wicke, Beusch, and Fieguth 1972). For the non-isothermal conditions they chose the oxidation of hydrogen by 8 mm spheres of silica alumina with 0·4 per cent platinum. As well as the transitions between steady states with the usual hysteresis effects, they observed a steady oscillation of period close to an hour at low gas flow rates. Figure 8.6 shows the measured temperatures at the centre (T_c) and surface (T_s) of the catalyst particle when the temperature of the gas stream was constant at $T_f = 70°C$. The reaction rate as measured by the concentration of water vapour in the product was also measured and is in phase with the temperature. It was found that with increasing temperature T_f the amplitude decreases and a stable steady state is reached. The oscillations begin spontaneously and are quite stable; the superposition of oscillations, resulting in double peaks, was also observed. The character of the oscillations is quite different from that of the limit cycle calculations, for the latter are sharply peaked in temperature with the greater part of the period spent at a low temperature. This suggests that a more subtle process connected with the mechanism of the reaction is at work.

Their work under isothermal conditions concerned the oxidation of carbon monoxide over palladium and platinum. Here the strong adsorption of

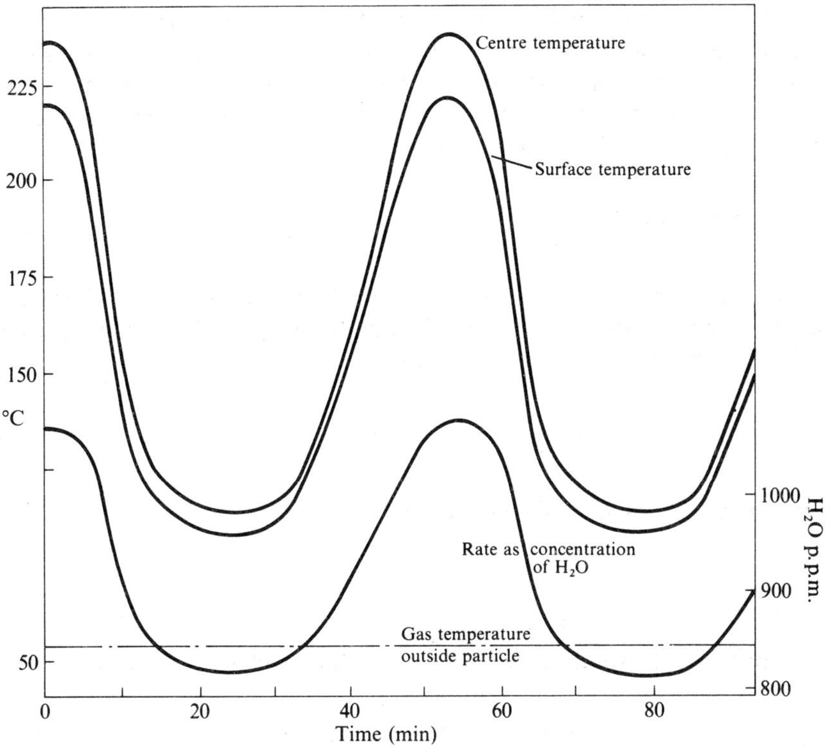

FIG. 8.6. Observed oscillations of temperature and reaction rate in a single catalyst particle for constant gas temperature. (After Wicke, Beusch, and Fieguth (1972).)

carbon monoxide inhibits the reaction at higher partial pressures as with the kinetic expression

$$R(u) = \frac{(1+\kappa)^2 u}{(\kappa+u)^2}.$$

In this case the oscillations of reaction rate are much more rapid with periods of the order of 2 min and an unsymmetrical form. Under certain conditions there are even more rapid flickerings of an almost random character. These tend to grow in amplitude, and at fairly regular intervals the reaction is virtually quenched for a few seconds and then resumes its former rate and fluctuation. Again this would seem to require a more detailed model of the mechanism than is afforded by such equations as (8.38) and (8.39).

8.3. Wei's bound for the maximum transient temperature

We have seen repeatedly that the temperature can rise well above the Prater temperature during the course of a transient and that this is particularly noticeable when the Lewis number is small. Wei has shown that it is possible to imagine kinetics that would produce an arbitrarily large temperature in an arbitrarily small region (Wei (1966)) and has given bounds for the maximum temperature. For the average temperature in the whole region Ω, Wei's result for the Dirichlet problem can be written

$$\langle v\rangle \leqslant \max\left\{1+\beta,\, 1+\frac{\beta}{\mathscr{L}},\, \left\langle v_0+\frac{\beta}{\mathscr{L}}u_0\right\rangle,\, \left\langle v_0+\frac{\beta}{\mathscr{L}}u_0\right\rangle+\frac{\beta(1-\mathscr{L})}{\mathscr{L}}\right\}. \quad (8.48)$$

To obtain this we start again from eqns (8.38) and (8.39) with the Dirichlet boundary conditions $u_s = v_s = 1$. Then multiplying the first equation by β and the second by $\mathscr{L}^{-1}$ and adding gives

$$\frac{\partial}{\partial\tau}(\beta u+v) = \frac{1}{\mathscr{L}}\nabla^2(\beta u+v)+\beta\left(1-\frac{1}{\mathscr{L}}\right)\frac{\partial u}{\partial\tau}. \quad (8.49)$$

Let $\mathscr{G}(\boldsymbol{\rho}, \boldsymbol{\sigma}, \tau)$ be the Green's function for the equation

$$\mathscr{L}\frac{\partial w}{\partial\tau} = \nabla^2 w \quad (8.50)$$

in Ω, and w be the solution of this equation for which $w_s = 1+\beta$ and

$$w(\boldsymbol{\rho}, 0) = w_0(\boldsymbol{\rho}) = \frac{\beta}{\mathscr{L}}u_0(\boldsymbol{\rho})+v_0(\boldsymbol{\rho}). \quad (8.51)$$

Then

$$w(\boldsymbol{\rho}, \tau) = \iiint\limits_{\Omega} \mathscr{G}(\boldsymbol{\rho}, \boldsymbol{\sigma}, \tau)w_0(\boldsymbol{\sigma})\, \mathrm{d}\Upsilon_\sigma - $$
$$- \int_0^\tau \mathrm{d}\tau' \iint\limits_{\partial\Omega} \frac{\partial}{\partial\nu}\mathscr{G}(\boldsymbol{\rho}, \boldsymbol{\sigma}, \tau-\tau')w_s\, \mathrm{d}\Sigma. \quad (8.52)$$

Also the solution of eqn (8.49) can be represented in the form

$$\beta u(\boldsymbol{\rho}, \tau)+v(\boldsymbol{\rho}, \tau) = \iiint\limits_{\Omega} \mathscr{G}(\boldsymbol{\rho}, \boldsymbol{\sigma}, \tau)(\beta u_0+v_0)\, \mathrm{d}\Upsilon +$$
$$+\beta\left(1-\frac{1}{\mathscr{L}}\right)\int_0^\tau \mathrm{d}\tau' \iiint\limits_{\Omega} \mathscr{G}(\boldsymbol{\rho}, \boldsymbol{\sigma}, \tau-\tau')\frac{\partial u}{\partial\tau'}\, \mathrm{d}\Upsilon -$$
$$- \int_0^\tau \mathrm{d}\tau' \iint\limits_{\partial\Omega} \frac{\partial}{\partial\nu}\mathscr{G}(\boldsymbol{\rho}, \boldsymbol{\sigma}, \tau-\tau')(\beta u_s+v_s)\, \mathrm{d}\Sigma. \quad (8.53)$$

The second term can be integrated by parts to give

$$\beta\left(1-\frac{1}{\mathscr{L}}\right)\left\{\iiint_{\Omega}\mathscr{G}(\boldsymbol{\rho},\boldsymbol{\sigma},0)u(\boldsymbol{\sigma},\tau)\,\mathrm{d}\Upsilon-\iiint_{\Omega}\mathscr{G}(\boldsymbol{\rho},\boldsymbol{\sigma},\tau)u_0(\boldsymbol{\sigma})\,\mathrm{d}\Upsilon\right\}-$$

$$-\beta\left(1-\frac{1}{\mathscr{L}}\right)\int_0^{\tau}\mathrm{d}\tau'\iiint_{\Omega}\frac{\partial\mathscr{G}}{\partial\tau'}u\,\mathrm{d}\Upsilon. \tag{8.54}$$

The first integral gives $\beta(1-1/\mathscr{L})u(\boldsymbol{\rho},\tau)$, since $\mathscr{G}(\boldsymbol{\rho},\boldsymbol{\sigma},0)$ is a Dirac measure, and may be taken to the other side of eqn (8.53); the second integral in (8.54) combines with the first and third of eqn (8.53) to give $w(\boldsymbol{\rho},\tau)$. Hence

$$\frac{\beta}{\mathscr{L}}u(\boldsymbol{\rho},\tau)+v(\boldsymbol{\rho},\tau)=w(\boldsymbol{\rho},\tau)-\beta\left(1-\frac{1}{\mathscr{L}}\right)\int_0^{\tau}\mathrm{d}\tau'\iiint_{\Omega}\frac{\partial\mathscr{G}}{\partial\tau'}u\,\mathrm{d}\Upsilon, \tag{8.55}$$

and taking the average over Ω gives

$$\langle v\rangle\leqslant\left\langle\frac{\beta}{\mathscr{L}}u+v\right\rangle=\langle w\rangle-\beta\left(1-\frac{1}{\mathscr{L}}\right)\left\langle\int_0^{\tau}\mathrm{d}\tau'\iiint_{\Omega}\frac{\partial\mathscr{G}}{\partial\tau'}u\,\mathrm{d}\Upsilon\right\rangle.$$

It may be shown that the last average takes its extreme values when u is either 0 or 1 and that

$$0\leqslant\int_0^{\tau}\mathrm{d}\tau'\left\langle\iiint_{\Omega}\frac{\partial\mathscr{G}}{\partial\tau'}\,\mathrm{d}\Upsilon\right\rangle\leqslant 1.$$

It follows that we should take $u=0$ when $\mathscr{L}>1$ and $u=1$ when $\mathscr{L}<1$ to get a safe estimate. Moreover since $\langle w\rangle$ cannot exceed $\langle w_0\rangle$ or $w_s=1+\beta$, whichever is the greater, we have the four possibilities given in eqn (8.48). These estimates are independent of the kinetics. Georgakis has shown (Aris and Georgakis, 1973) that the bound on $\langle v\rangle$ for the Robin problem analogous to eqn (8.48) only requires that the terms $1+(\nu\beta/\mu)$ and $1+(\nu\beta/\mu)+\beta(1-\mathscr{L})/\mathscr{L}$ be added to the set from which the greatest is to be chosen.

Wei gives an extreme illustration by pointing out that a unit pulse of heat initiated over the plane ρ' at time τ' causes the temperature at the plane $\rho=0$ to reach a maximum value of $(2\pi e)^{-\frac{1}{2}}(\rho')^{-1}$ at time $\tau'+\frac{1}{2}(\rho')^2$. By initiating such a pulse over the plane ρ' at a time $-\frac{1}{2}(\rho')^2$, $0\leqslant\rho'\leqslant 1$, the maxima would reach the plane $\rho=0$ simultaneously at $\tau=0$ and produce a maximum temperature proportional to $\int_0^1(\rho')^{-1}\,\mathrm{d}\rho'$, which is infinite. It is of course unrealistic to consider such a concentration of heat but it does suggest that though the average temperature is bounded, the local fluctuations may be very large within a confined volume. To estimate a bound for the temperature in a small region, Wei used the following ingenious idea. If the space variables in Ω are discretized the mesh size becomes a measure of the small region under scrutiny, and the equations become ordinary differential equations. If $\mathbf{u}$ and $\mathbf{v}$ are vectors of concentrations $u_i(\tau)$ and temperatures $v_i(\tau)$ at the mesh points the discrete form of the equations is

$$\frac{\mathrm{d}\mathbf{u}}{\mathrm{d}\tau} = \mathbf{A}\mathbf{u} - \phi^2\mathbf{r},$$

$$\mathscr{L}\frac{\mathrm{d}\mathbf{v}}{\mathrm{d}\tau} = \mathbf{A}\mathbf{v} + \beta\phi^2\mathbf{r},$$

and these can be combined as before to give

$$\frac{\mathrm{d}}{\mathrm{d}\tau}(\beta\mathbf{u}+\mathbf{v}) = \frac{1}{\mathscr{L}}\mathbf{A}(\beta\mathbf{u}+\mathbf{v}) + \beta\left(1-\frac{1}{\mathscr{L}}\right)\frac{\partial\mathbf{u}}{\partial\tau}.$$

This leads by similar manipulations to a discrete analogue of eqn (8.55)

$$\frac{\beta}{\mathscr{L}}\mathbf{u}+\mathbf{v} = \mathbf{w} - \beta\left(1-\frac{1}{\mathscr{L}}\right)\int_0^\tau \frac{\mathrm{d}\mathbf{G}}{\mathrm{d}\tau'}\mathbf{u}\,\mathrm{d}\tau', \tag{8.56}$$

where

$$\mathbf{G} = \exp\left(\mathbf{A}\frac{\tau-\tau'}{\mathscr{L}}\right)$$

and

$$\mathbf{w} = [\exp(\mathbf{A}\tau/\mathscr{L})]\mathbf{w}_0.$$

Wei was able to establish a bound on the integral for the geometry of a slab sphere. It took the form of showing that the largest term would be bounded by $a+b\log N$, where N is the number of points in the discretization, $a = 0{\cdot}9365$, $b = 1{\cdot}1145$ for the slab and $a = 0{\cdot}529$, $b = 2{\cdot}130$ for the sphere. It seems likely that his results could be generalized to an arbitrary shape and they would then give

$$\max_{\tau\geqslant 0}\max_{1\leqslant i\leqslant N}[v_i(\tau)] \leqslant \max_{1\leqslant i\leqslant N}[w_{0i}(0)] + \beta\frac{1-\mathscr{L}}{\mathscr{L}}(a+b\log N). \tag{8.57}$$

The number N may be interpreted as proportional to the ratio of the volume of Ω to that of the small volume under consideration, and this shows that the bound on the temperature increases logarithmically with this ratio of volumes.

8.4. Singular perturbation analysis

The way in which the analogy of the stirred tank can be more rigorously justified is to suppose that the diffusion of heat and matters is very rapid compared with the reaction. On physical grounds this makes sense for letting

D_e and k_e go to infinity, accomplishing the perfect mixing that is characteristic of the stirred tank. This is the line of approach that Cohen (1973) and Poore (1972, 1973) have taken in looking at the tubular reactor (see also Cohen and Poore (1973)). We will introduce it here in connection with the catalytic body and look at its bearing on the results of Section 8.2.

In eqns (8.38–40) we propose to let ϕ, μ and ν tend to zero while keeping

$$\nu' = \nu/\phi^2 \quad \text{and} \quad \mu' = \mu/\phi^2 \tag{8.58}$$

finite.

Since the diffusion coefficient enters the definition of the dimensionless time τ, a new dimensionless time

$$\tau' = \phi^2\tau = \rho_b S_g \hat{r}(c_f, T_f)t/c_f \tag{8.59}$$

will be defined. Let u and v be written as series

$$u(\mathbf{\rho}, \tau') = \sum_{n=0}^{\infty} \phi^{2n} u_n(\mathbf{\rho}, \tau'),$$

$$v(\mathbf{\rho}, \tau') = \sum_{n=0}^{\infty} \phi^{2n} v_n(\mathbf{\rho}, \tau') \tag{8.60}$$

and substituted in the equations. Then collecting terms of the same order gives:

$$\nabla^2 u_0 = 0, \qquad \nabla^2 v_0 = 0, \tag{8.61}$$

$$\frac{\partial u_0}{\partial \nu} = 0, \qquad \frac{\partial v_0}{\partial \nu} = 0, \tag{8.62}$$

$$\frac{\partial u_0}{\partial \tau'} = \nabla^2 u_1 - \phi^2 R(u_0, v_0), \qquad \mathscr{L}\frac{\partial v_0}{\partial \tau'} = \nabla^2 v_1 + \beta\phi^2 R(u_0, v_0), \tag{8.63}$$

$$\frac{1}{\nu'}\frac{\partial u_1}{\partial \nu} = 1 - u_0, \qquad \frac{1}{\mu'}\frac{\partial v_1}{\partial \nu} = 1 - v_0, \tag{8.64}$$

$$\frac{\partial u_1}{\partial \tau'} = \nabla^2 u_2 - \phi^2\{(R_u)_0 u_1 + (R_v)_0 v_1\},$$

$$\mathscr{L}\frac{\partial v_1}{\partial \tau'} = \nabla^2 v_2 + \beta\phi^2\{(R_u)_0 u_1 + (R_v)_0 v_1\} \tag{8.65}$$

$$\frac{1}{\nu'}\frac{\partial u_2}{\partial \nu} = -u_1, \qquad \frac{1}{\mu'}\frac{\partial v_2}{\partial \nu} = -v_1. \tag{8.66}$$

The first equations, (8.60) and (8.61), show that

$$u_0(\mathbf{\rho}, \tau') = \bar{u}(\tau'), \qquad v_0(\mathbf{\rho}, \tau') = \bar{v}(\tau') \tag{8.67}$$

must in fact be independent of position, but they do not provide equations for these time-varying functions. These come from averaging eqns (8.63) and using the boundary condition (8.64). For if u_0 is independent of position it equals its average value and

$$\frac{d\bar{u}}{d\tau'} = \frac{1}{\upsilon}\iiint_{\Omega} \nabla^2 u_1 \, d\gamma - \phi^2 R(\bar{u}, \bar{v})$$

$$= \frac{1}{\upsilon}\iint_{\partial\Omega} \frac{\partial u_1}{\partial v} \, d\Sigma - \phi^2 R(\bar{u}, \bar{v})$$

$$= \frac{v'\sigma}{\upsilon}(1-\bar{u}) - \phi^2 R(\bar{u}, \bar{v}). \tag{8.68}$$

Similarly

$$\mathscr{L}\frac{d\bar{v}}{d\tau'} = \frac{\mu'\sigma}{\upsilon}(1-\bar{v}) + \beta\phi^2 R(\bar{u}, \bar{v}). \tag{8.69}$$

Equations (8.68) and (8.69) are equivalent to (8.41) and (8.42) if τ' is replaced by $(v'\sigma\tau'/\upsilon) = (v\sigma\tau/\upsilon)$ and $\psi^2 = (\phi^2\upsilon/v'\sigma)$, $\beta' = (v'\beta/\mu')$, $L = (\mathscr{L}v'/\mu')$.

The full transient equations are subject to initial conditions, which cannot be satisfied by these ordinary differential equations. To obtain initial values for $\bar{u}$ and $\bar{v}$ we have to use an expanded scale of time for the beginning of the process. Thus we take an expanded scale of time τ'/ϕ^2 which is of course the old time variable τ. The zeroth-order terms in the expansion of this inner solution will satisfy the original equations with $\phi^2 = \mu = v = 0$, i.e.

$$\frac{\partial U_0}{\partial \tau} = \nabla^2 U_0, \qquad \mathscr{L}\frac{\partial V_0}{\partial \tau} = \nabla^2 V_0 \quad \text{in } \Omega \tag{8.70}$$

with

$$\frac{\partial U_0}{\partial v} = 0, \qquad \frac{\partial V_0}{\partial v} = 0 \quad \text{on } \partial\Omega$$

and

$$U_0(\boldsymbol{\rho}, \tau) = u_0(\boldsymbol{\rho}), \qquad V_0(\boldsymbol{\rho}, \tau) = v_0(\boldsymbol{\rho}).$$

By the matching principle for singular perturbations $\bar{u}$ and $\bar{v}$ at $\tau' = 0$ are given by the values of U_0 and V_0 as $\tau \to \infty$. But eqns (8.70) just represent the process of diffusion in an isolated region Ω so that the limiting values are constant and equal to the mean values of the original distributions, that is

$$\bar{u}(0) = \langle u_0 \rangle, \qquad \bar{v}(0) = \langle v_0 \rangle. \tag{8.71}$$

Thus for large values of k_e and D_e the concentration and temperature quickly become uniform and then are governed by the stirred tank equations.

If $(\bar{u}_e, \bar{v}_e)$ is a steady state of the stirred tank and $u^* = \bar{u}-u_e$, $v^* = \bar{v}-v_e$ we can write

$$\frac{du^*}{d\tau'} = -\{1+\psi^2(R_u)_e\}u^* - \psi^2(R_v)_e v^* - \psi^2 S(u^*, v^*), \tag{8.72}$$

$$\frac{dv^*}{d\tau'} = \frac{\beta'\psi^2}{L}(R_u)_e u^* - \frac{1}{L}\{1-\beta'\psi^2(R_v)_e\} + \frac{\beta'\psi^2}{L}S(u^*, v^*), \tag{8.73}$$

where

$$S(u^*, v^*) = R(u_e+u^*, v_e+v^*) - R(u_e, v_e) - (R_u)_e u^* - (R_v)_e v^*, \tag{8.74}$$

is a function whose value and first derivatives all vanish at $u^* = v^* = 0$. By the usual linearization technique we see that the steady state $(\bar{u}_e, \bar{v}_e)$ will be stable if

$$\{1+\psi^2(R_u)_e\} + \frac{1}{L}\{1-\beta'\psi^2(R_v)_e\} > 0 \tag{8.75}$$

and

$$1+\psi^2\{(R_u)_e - \beta'(R_v)_e\} > 0. \tag{8.76}$$

The second condition is not affected by the value of L, but if $\beta'\psi^2(R_v)_e < 1$, the first condition can be violated by taking L to be sufficiently small.

Let $\lambda = L^{-1}$ and

$$\lambda_c = \{1+\psi^2(R_u)_e\}/\{\beta'\psi^2(R_v)_e - 1\} \tag{8.77}$$

be positive then if (8.76) is satisfied, then the steady state will be stable if $\lambda < \lambda_c$, and unstable if $\lambda > \lambda_c$. By a linear transformation of the dependent variable to

$$\begin{aligned} x(\tau') &= a(\lambda)u^*(\tau') + b(\lambda)v^*(\tau'), \\ y(\tau') &= c(\lambda)u^*(\tau') + d(\lambda)v^*(\tau'), \end{aligned} \tag{8.78}$$

the equations can be brought into the form

$$\begin{aligned} \frac{dx}{d\tau'} &= \alpha(\lambda)x - \beta(\lambda)y + \gamma(\lambda)T(x, y) \\ \frac{dy}{d\tau'} &= \beta(\lambda)x + \alpha(\lambda)y + \delta(\lambda)T(x, y) \end{aligned} \tag{8.79}$$

where

$$T(x, y) = S(u^*, v^*). \tag{8.80}$$

Moreover, since the stability characteristics are not thereby altered, the transition to stability by the vanishing of the real part of the eigenvalues must occur at $\lambda = \lambda_c$, that is

$$\alpha(\lambda_c) = 0. \tag{8.81}$$

The linearized phase portrait in the neighbourhood of the steady state when $\lambda = \lambda_c$ is given by

$$\frac{dx}{d\tau'} = -\beta y, \qquad \frac{dy}{d\tau'} = \beta x,$$

so that the trajectories would be circles were it not for the non-linear terms which act like a forcing oscillation. Now the system of equations

$$\begin{aligned} \dot{x} + \beta y &= m \sin \beta t + n \cos \beta t \\ \dot{y} - \beta x &= p \sin \beta t + q \cos \beta t \end{aligned} \tag{8.82}$$

will only have bounded solutions if $m = q$ and $n = -p$ and the idea behind Cohen's analysis is that the approximation should be built up with bounded solutions at each stage. Two time scales are introduced, one of which is deliberately slow in the neighbourhood of $\lambda = \lambda_c$. Let $\varepsilon = \lambda - \lambda_c$ and define

$$t = \varepsilon^2 \tau', \qquad T = (1 + \varepsilon\omega_1 + \varepsilon^2\omega_2 + \dots)\tau', \tag{8.83}$$

putting

$$\begin{aligned} x &= \varepsilon x_1(t, T) + \varepsilon^2 x_2(t, T) + \dots \\ y &= \varepsilon y_1(t, T) + \varepsilon^2 y_2(t, T) + \dots \end{aligned} \tag{8.84}$$

in eqns (8.79). When these are expanded in powers of ε it is found that x_1 and y_1 satisfy

$$\frac{\partial x_1}{\partial T} + \beta y_1 = 0, \qquad \frac{\partial y_1}{\partial T} - \beta x_1 = 0,$$

so that x_1 and y_1 are oscillations in fast time, T, with amplitudes that grow in slow time, t:

$$\begin{aligned} x_1 &= A(t) \sin \beta T + B(t) \cos \beta T \\ y_1 &= B(t) \sin \beta T - A(t) \cos \beta T. \end{aligned}$$

By insisting on a bounded solution for x_2 and y_2 we find that $\omega_1 = 0$. However an equation for $A(t)$ and $B(t)$ does not arise until the boundedness of the solution for x_3 and y_3 is enforced. This is because

$$\frac{d}{d\tau'} = (1 + \varepsilon\omega_1 + \varepsilon^2\omega_2 + \dots)\frac{\partial}{\partial T} + \varepsilon^2 \frac{\partial}{\partial t},$$

so that the first entry of a derivative with respect to slow time is with the terms of order ε^2. The algebraic labour is considerable but this results in an equation for $C(t)$, the square of the amplitude, $A^2(t)+B^2(t)$. This equation has the form

$$\frac{dC}{dt} = PC - QC^2,$$

where P and Q are constants that can be evaluated. This equation shows that C grows to the value P/Q if P and Q are positive, so that there is a limit cycle of amplitude $(P/Q)^{\frac{1}{2}}\varepsilon^{\frac{1}{2}}$ for $\varepsilon > 0$.

This type of analysis is extremely laborious but it does show how far the stirred tank analogy can be taken in seeking simple ways to illuminate the behaviour of the catalyst particle.

8.5. Periodic operation

There has been quite an interest in recent years in the possibility of deliberately operating chemical reactors in a cyclic manner under such conditions that the average performance over a cycle is preferable to any steady state operation. We shall consider only the aspect of this effort that has so far led to problems involving diffusion and reaction. This is to be found in the work of Horn and Bailey (1968, 1969, 1970, 1972) who used the following example. Two products B_1 and B_2 are formed from a reactant A_1. The first is formed by catalytic action in two steps: adsorption on or complexing with a catalytic site S to form the complex A_2, followed by a further reaction between A_1 and the adsorbed molecule A_2 to form B_1. The second product is formed directly from A_1 by a second order reaction. Thus we have the reaction system

$$A_1 + S \underset{2}{\overset{1}{\rightleftharpoons}} A_2,$$

$$A_1 + A_2 \xrightarrow{3} 2B_1 + S,$$

$$2A_1 \xrightarrow{4} 2B_2.$$

If the surface coverage is low, so that the concentration of sites does not limit the reaction, the rates per unit catalyst area of the four steps can be written $\hat{k}_1 c_1$, $\hat{k}_2 c_2$, $\hat{k}_3 c_1 c_2$ and $\hat{k}_4 c_1^2$ respectively, where c_1 and c_2 are the concentrations of A_1 and A_2. If the complex A_2 is immobilized on the catalyst it does not diffuse and mass balances on A_1 and A_2 give

$$\varepsilon \frac{\partial c_1}{\partial t} = D_e \nabla^2 c_1 + \rho_b S_g(-\hat{k}_1 c_1 + \hat{k}_2 c_2 - \hat{k}_3 c_1 c_2 - 2\hat{k}_4 c_1^2) \tag{8.85}$$

$$\varepsilon \frac{\partial c_2}{\partial t} = \rho_b S_g(\hat{k}_1 c_1 - \hat{k}_2 c_2 - \hat{k}_3 c_1 c_2) \tag{8.86}$$

in Ω, with $c_1 = c_{1s}$ on $\partial\Omega$. Since c_{1s} is to be varied we will not use it to make the concentrations dimensionless but choose instead $c_f = \max c_{1f}$ setting

$$\begin{aligned} u_1 &= c_1/c_f, \quad u_2 = c_2/c_f, \quad \phi^2 = a^2 \rho_b S_g \hat{k}_3 c_f / D_e, \\ \tau &= D_e t / a^2 \varepsilon, \quad \kappa = \hat{k}_2 / \hat{k}_3 c_f, \quad \lambda = \hat{k}_1 / \hat{k}_3 c_f, \quad \mu = 2\hat{k}_4 / \hat{k}_3. \end{aligned} \tag{8.87}$$

Then

$$\frac{\partial u_1}{\partial \tau} = \nabla^2 u_1 + \phi^2 (\kappa u_2 - \lambda u_1 - \mu u_1^2 - u_1 u_2), \tag{8.88}$$

$$\frac{\partial u_2}{\partial \tau} = \phi^2 \{\lambda u_1 - (\kappa + u_1) u_2\} \tag{8.89}$$

in Ω, with

$$u_1 = u_{1s}(\tau) \tag{8.90}$$

on $\partial\Omega$. Also we consider that u_{1s} is a periodic function of period T, i.e.

$$u_{1s}(\tau + T) = u_{1s}(\tau). \tag{8.91}$$

If equations for the products B_1 and B_2 were written down they would be

$$\varepsilon \frac{\partial b_1}{\partial t} = D_{1e} \nabla^2 b_1 + 2\rho_b S_g \hat{k}_3 c_1 c_2,$$

$$\varepsilon \frac{\partial b_2}{\partial t} = D_{2e} \nabla^2 b_2 + 2\rho_b S_g \hat{k}_4 c_1^2,$$

or, in dimensionless form,

$$\frac{\partial v_1}{\partial \tau} = \delta_1 \nabla^2 v_1 + 2\phi^2 u_1 u_2,$$

$$\frac{\partial v_2}{\partial \tau} = \delta_2 \nabla^2 v_2 + \mu \phi^2 u_1^2.$$

In the established operation of the cyclic processes, all the dependent variables will be periodic functions of τ of period T, hence the rate of production of B_1 over a period will be proportional to

$$\begin{aligned} -\frac{1}{T} \int_0^T \mathrm{d}\tau \iint_{\partial\Omega} \delta_1 \frac{\partial v_1}{\partial \nu} \mathrm{d}\Sigma &= -\frac{1}{T} \int_0^T \mathrm{d}\tau \iiint_{\Omega} \delta_1 \nabla^2 v_1 \mathrm{d}\Upsilon \\ &= \frac{1}{T} \int_0^T \mathrm{d}\tau \iiint_{\Omega} \left[2\phi^2 u_1 u_2 - \frac{\partial v_1}{\partial \tau} \right] \mathrm{d}\Upsilon \\ &= \frac{2\phi^2}{T} \int_0^T \mathrm{d}\tau \iiint_{\Omega} u_1 u_2 \, \mathrm{d}\Upsilon, \end{aligned}$$

since the periodicity ensures the vanishing of the integral of $\partial v_1/\partial\tau$. As dimensionless production rates we can write

$$\begin{aligned} P_1 &= 2\phi^2 \frac{1}{\upsilon T} \int_0^T \mathrm{d}\tau \iiint_\Omega u_1 u_2 \, \mathrm{d}\Upsilon \\ P_2 &= \mu\phi^2 \frac{1}{\upsilon T} \int_0^T \mathrm{d}\tau \iiint_\Omega u_1^2 \, \mathrm{d}\Upsilon. \end{aligned} \tag{8.92}$$

In steady state operation the complex, which does not diffuse, satisfies the pseudo-steady state hypothesis (cf. Section 5.6) for, setting the time derivative in eqn (8.89) equal to zero we have

$$u_2 = \lambda u_1/(\kappa + u_1) \tag{8.93}$$

and hence

$$\nabla^2 u_1 = \phi^2 \left\{ \frac{2\lambda}{\kappa + u_1} + \mu \right\} u_1^2. \tag{8.94}$$

For steady state operation with $u_1 = u_{1s}$, a constant value of the surface concentration, this equation in Ω would be solved with $u_1 = u_{1s}$ on $\partial\Omega$ and an effectiveness factor

$$\eta = \eta(\phi\,;\kappa, \lambda, \mu\,; u_{1s}) = \left[\iint_{\partial\Omega} \left(\frac{\partial u_1}{\partial v} \right) \mathrm{d}\Sigma \right] \bigg/ \left[\phi^2 u_{1s}^2 \left(\frac{2\lambda}{\kappa + u_{1s}} + \mu \right) \right] \tag{8.95}$$

for the consumption of the reactant could be defined. We note in passing that $M(u_{1s}; 0)$, the factor for normalizing the Thiele modulus, is given by

$$M(u_{1s}; 0) = \frac{(2\lambda[u_{1s}^2 - 2\kappa u_{1s} + 2\kappa^2 \ln\{(\kappa + u_{1s})/\kappa\}] + \frac{2}{3}\mu u_{1s}^3)^{\frac{1}{2}}}{u_{1s}^2\{2\lambda/(\kappa + u_{1s}) + \mu\}} \tag{8.96}$$

and that then

$$\eta\Phi \sim 1, \qquad \Phi = \phi/M(u_{1s}; 0) \tag{8.97}$$

for large ϕ or Φ.

Consider first the situation when there is no diffusion limitation so that $u_1 \equiv u_{1s}$, the prescribed periodic concentration, and u_2 is given by eqn (8.89). If u_{1s} is constant the production rates are

$$P_1 = 2\phi^2\lambda u_{1s}^2/(\kappa + u_{1s}) \quad \text{and} \quad P_2 = \mu\phi^2 u_{1s}^2$$

respectively, and since by the normalization u_{1s} may range from 0 to 1, the entire range of steady state performance may be represented by the curve S in Fig. 8.7. The point O corresponds to $u_{1s} = 0$ and A to $u_{1s} = 1$. But any point in the convex hull of S can also be attained by slow cycling. For, consider a long period rectangular wave in which $u_{1s} = 0$, $0 < \tau < \alpha T$,

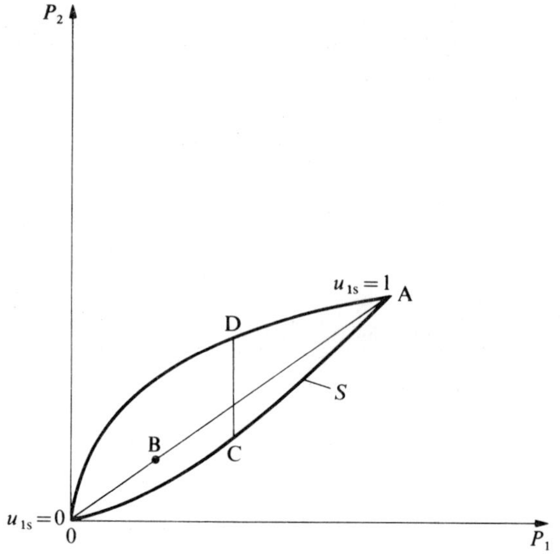

FIG. 8.7. The region of accessible production rates with no diffusion.

$u_{1s} = 1$, $\alpha T < \tau < T$. The transients in passing from one steady state to another could be made negligibly small by taking T sufficiently long, and the mixed product would then correspond to the production rates at a point B on OA with AB $= \alpha$OA. By slow cycling between two values of u_{1s} any point in the area between OA and S, i.e. the convex hull of S, can be attained.

The opposite scheme of switching very rapidly between two values of u would be realized by putting

$$u_{1s}(\tau) = \begin{cases} U^* & 0 < \tau < \alpha T, \\ U_* & \alpha T < \tau < T, \end{cases} \tag{8.98}$$

and letting $T \to 0$. Because eqn (8.89) happens to be linear in u_2 it can be solved explicitly and the calculation leads to an average value

$$\begin{aligned} \langle u_2 \rangle &= \lambda \langle u_1 \rangle / (\kappa + \langle u_1 \rangle), \\ \langle u_1 \rangle &= \alpha U^* + (1-\alpha) U_*, \end{aligned} \tag{8.99}$$

and average production rates

$$\begin{aligned} P_1 &= 2\phi^2 \lambda \langle u_1 \rangle^2 / (\kappa + \langle u_1 \rangle), \\ P_2 &= \mu \phi^2 \{ \langle u_1 \rangle^2 + \alpha(1-\alpha)(U^* - U_*)^2 \}. \end{aligned} \tag{8.100}$$

Now if $\alpha U^* + (1-\alpha) U_* = \langle u_1 \rangle$, the second term in the expression for P_2 is $(U^* - \langle u_1 \rangle)(\langle u_1 \rangle - U_*)$ and this can be as large as $(1 - \langle u_1 \rangle)\langle u_1 \rangle$. Hence, if in Fig. 8.7 the point C on the steady state curve S corresponds to $u_{1s} = \langle u_1 \rangle$, production rates of B_2 can be found corresponding to any point of the line CD by suitably choosing α, U^* and U_*. The cases of interest will arise when D rises above the chord OA, for now there are relative production rates which are attainable by dynamic operation which cannot be reached by slow cycling, namely those corresponding to points in the convex hull of the curve ODA. In fact this upper curve is given by

$$P_1 = 2\phi^2\langle u_1\rangle^2/(\kappa+\langle u_1\rangle), \qquad P_2 = \mu\phi^2\langle u_1\rangle \tag{8.101}$$

so that its tangent at the origin is vertical and any ratio P_2/P_1 may be obtained by rapid fluctuations in the neighbourhood of the origin.

Needless to say the computation would be very much more difficult if the equation for u_2 were not linear. Horn and Bailey appeal to a theorem of Warga (1962), which shows that such rapid fluctuations are a way of approximating the solution to a relaxed control problem. The production rates of B_1 and B_2 are indeed given at a 'relaxed steady state.' The steady state production rates were found by putting $(\partial u_2/\partial\tau) = 0$ in the equations

$$P_1 = \frac{\partial v_1}{\partial \tau} = 2\phi^2 u_1 u_2 = f_1(u_1, u_2),$$

$$P_2 = \frac{\partial v_2}{\partial \tau} = \mu\phi^2 u_1^2 = f_2(u_1, u_2),$$

$$0 = \frac{\partial u_2}{\partial \tau} = \phi^2\{\lambda u_1 - (\kappa+u_1)u_2\} = f_3(u_1, u_2),$$

or by saying that the production vector (P_1, P_2) should be such that $(P_1, P_2, 0) \in \mathrm{S}$ where S is the set of all vectors (f_1, f_2, f_3) and u_1 ranges over its permitted interval (0, 1). In the relaxed steady state the vector $(P_1, P_2, 0)$ is only required to lie in the convex hull of (f_1, f_2, f_3). In this case we would find the surface in f-space, for which u_1 and u_2 are parameters, form its convex hull, and take the section of the plane of P_1 and P_2. This is the lenticular area ODACO of Fig. 8.7. Horn and Bailey (1968) show how it may be framed as a programming problem.

If the selectivity can be varied by cyclic operation when there is no diffusion limitation, there is some possibility that it may also be of value in the catalyst pellet. In this case however we notice that there is not only a time constant for the complex-formation or adsorbed intermediate, but also one for the diffusion process itself. We might suspect that that might be an optimal frequency for the cycling, since too rapid a change at the surface will not penetrate into the interior. We will restrict $u_{1s}(\tau)$ to a rectangular wave of maximum amplitude and write

$$u_{1s}(\tau) = U_\alpha(\tau) = \begin{Bmatrix} 1 & 0 \leqslant \tau < \alpha T, \\ 0 & \alpha T \leqslant \tau < T \end{Bmatrix} = U_\alpha(\tau + T), \tag{8.102}$$

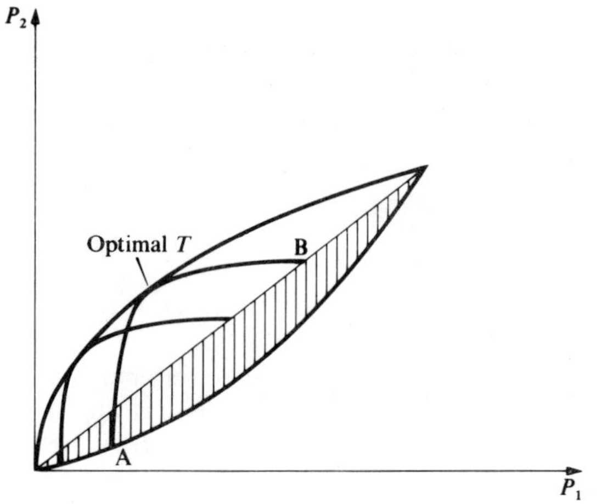

FIG. 8.8. The region of steady states and its convex hull with several loci of average production rates under periodic operation.

where $0 \leqslant \alpha \leqslant 1$. For a given value of the Thiele modulus and a given set of kinetic constants, the steady state production rates can be calculated and their convex hull drawn in the (P_1, P_2)-plane: this is the shaded region in Fig. 8.8. For fixed α the average production rates given by eqn (8.92) may be calculated for the surface fluctuations of eqn (8.102). A point is obtained for each value, giving a curve (AB in Fig. 8.8) which runs from a steady state point when $T \to 0$ to a point on the convex hull as $T \to \infty$. The attainable region under dynamic control is thus the envelope of all such curves as γ varies from 0 to 1. Horn and Bailey showed that these curves did indeed pass out of the convex hull of the steady state region. Moreover there is an optimal value of the period of oscillation T, corresponding to the value of T where the curve touches the envelope. Gains in selectivity over steady state operation were found even when the reaction was severely limited.

In a later study Bailey and Lee (1974) considered the same reaction system but made the first step irreversible and allowed the complex A_2 to diffuse. Thus in place of eqns (8.88) and (8.89) we would have

$$\frac{\partial u_1}{\partial \tau} = \nabla^2 u_1 - \phi^2 u_1(\lambda + \mu u_1 + u_2),$$

$$\frac{\partial u_2}{\partial \tau} = \delta \nabla^2 u_2 + \phi^2 u_1(\lambda - u_2),$$

where δ is the ratio of the effective diffusion coefficient for A_2 to D_e, the effective diffusion coefficient for A_1. As before, u_{1s} is a periodic function of τ while $u_{2s} = 0$ on $\partial\Omega$.

For small values of the Thiele modulus, ϕ, Bailey and Lee use a perturbation analysis with $\lambda\phi^2$ as the parameter:

$$u_i = u_{i0} + \lambda\phi^2 u_{i1} + \lambda^2\phi^4 u_{i2} + \dots, \quad i = 1, 2.$$

This is a regular perturbation and gives a sequence of equations for $u_{10}, u_{20}, u_{11}, u_{21} \dots$. The equation for u_{20} admits the periodic solution $u_{20} = 0$, so that the lowest terms in P_1 and P_2 are

$$P_1 = 2\lambda\phi^4 \frac{1}{\upsilon T} \int_0^T \mathrm{d}\tau \iiint_\Omega u_{10} u_{21} \, \mathrm{d}\Upsilon + \dots$$

and

$$P_2 = \mu\phi^2 \frac{1}{\upsilon T} \int_0^T \mathrm{d}\tau \iiint_\Omega u_{10}^2 \, \mathrm{d}\Upsilon + \dots$$

and for these only u_{10} and u_{21} need be determined. These satisfy

$$\frac{\partial u_{10}}{\partial \tau} = \nabla^2 u_{10}, \qquad u_{10} = u_{1s}(\tau),$$

$$\frac{\partial u_{21}}{\partial \tau} = \delta\nabla^2 u_{21} + u_{10}, \qquad u_{21} = 0.$$

To illustrate this technique Bailey considered Ω to be a slab and used a sinusoidal perturbation. Even for this case some quite complicated formulae arise, but the computed results show that a substantial gain in selectivity can be obtained by cycling at a finite rate. If σ/σ_∞ is the ratio of the selectivity when the period is T to that when $T \to \infty$, then the maxima found by Bailey with $\lambda = 1$, $\mu = 2$, and

$$u_{1s}(\tau) = \tfrac{1}{2}\left(1 + \cos 2\pi \frac{\tau}{T}\right)$$

are:

δ	$(\sigma/\sigma_\infty)_{\max}$	$T_{\max}$
0·5	1·23	2·85
1·0	1·16	1·86
2·0	1·11	1·15

Thus even in the case when diffusion is no great limitation, substantial changes in selectivity can be obtained.

To explore the effects when diffusion is a severe limitation, Bailey and Lee were forced to do direct numerical calculations. Here even more substantial changes in selectivity were found, as the following results for the same parameters but $\phi^2 = 10$ show:

δ	$(\sigma/\sigma_\infty)_{max}$	T_{max}
0·5	1·40	0·56
1·0	1·375	0·52
2·0	1·35	0·43

Thus the gain is greater but is achieved at somewhat higher frequencies. It should also be said that the maxima are rather flatter; for example, with $\delta = 0{\cdot}5$, σ/σ_∞ is greater than 1·35 for $0{\cdot}18 < T < 2{\cdot}18$.

8.6. An optimal control problem

Lions (1972) has discussed an optimal control problem which might arise in a biochemical context. An isothermal reaction can have its rate modified by an inhibiting species which also diffuses but does not react. In a chemical engineering context this might be a form of reversible poisoning of the catalyst. If u denotes the concentration of the reacting species and v that of the inhibitor we would have the usual equations

$$\frac{\partial u}{\partial \tau} = \nabla^2 u - \phi^2 R(u, v), \tag{8.103}$$

$$\frac{\partial v}{\partial \tau} = \delta \nabla^2 v \tag{8.104}$$

in Ω, δ being the ratio of the diffusivities of inhibitor and reactant. The boundary conditions must show where the control problem arises and where it can be solved. Thus a possible situation is that the concentration of reactant on $\partial\Omega$, u_s, varies, and we want to compensate by deliberately varying v_s. Then we would have

$$u_s = f(\mathbf{\rho}, \tau) \quad \text{on } \partial\Omega \tag{8.105}$$

and

$$v_s = V(\tau) \qquad \text{on } \partial\Omega, \tag{8.106}$$

where $V(\tau)$ is the control policy for the surface concentration of inhibitor. $V(\tau)$ would be subject to certain practical and necessary limitations such as a bound

$$0 \leqslant V(\tau) \leqslant V^*. \tag{8.107}$$

Since this is a time-dependent problem the initial conditions are needed and for simplicity these are

$$u(\mathbf{\rho}, 0) = v(\mathbf{\rho}, 0) = 0 \quad \text{in } \Omega. \tag{8.108}$$

The inhibitory character of v is evident if we put

$$R_u(u, v) > 0, \qquad R_v(u, v) < 0. \tag{8.109}$$

Finally we observe the concentration u at some internal point, which we can take to be the origin $\boldsymbol{\rho} = 0$, and try to find $V(\tau)$ so that this concentration will be as close to a premeditated value $U(\tau)$ as possible, that is we seek to minimize

$$J(V) = \int_0^T |u(0, \tau) - U(\tau)|^2 \, d\tau. \tag{8.110}$$

The equations as given should have a unique solution for any admissible control, and Lions in his article shows this for a particular form of $R(u, v)$. He also demonstrates the existence of the optimal control by showing that when V spans the admissible range $(0, V)$, $v(\boldsymbol{\rho}, \tau)$ remains in a relatively compact subset of $L^2(\Omega \times (0, \tau))$ and hence that the optimal $V(\tau)$, say $V^0(\tau)$, is an admissible control and that $J(V) \geqslant J(V^0)$ for all other controls. Whether or not V^0 is unique is an open question.

To obtain necessary conditions for optimality we consider a deviation from the optimal control and set

$$V(\tau) = V^0(\tau) + \varepsilon w(t), \tag{8.111}$$

where ε is sufficiently small that $V(\tau)$ is still admissible. Then if $V^0(\tau)$ is indeed optimal we shall have

$$J'(V^0) = \left\{ \frac{d}{d\varepsilon} J(V^0 + \varepsilon w) \right\}_{\varepsilon = 0} \geqslant 0 \tag{8.112}$$

for all w, since if V^0 is in the interior of the admissible control region it will be zero, whereas if it is on the boundary the derivative *into* the admissible space must be positive. Let $\dot{u}$ and $\dot{v}$ denote

$$\left\{ \frac{d}{d\varepsilon} u(\boldsymbol{\rho}, \tau; V^0 + \varepsilon w) \right\}_{\varepsilon = 0} \quad \text{and} \quad \left\{ \frac{d}{d\varepsilon} v(\boldsymbol{\rho}, \tau; V^0 + \varepsilon w) \right\}_{\varepsilon = 0},$$

respectively, where $u(\boldsymbol{\rho}, \tau; v_s)$ is an obvious notation for the dependence of the solutions of eqns (8.103–8) on the controlled boundary concentration v_s. Then $\dot{u}$ and $\dot{v}$ satisfy

$$\frac{\partial \dot{u}}{\partial \tau} = \nabla^2 \dot{u} - \phi^2 R_u \dot{u} - \phi^2 R_v \dot{v}, \tag{8.113}$$

$$\frac{\partial \dot{v}}{\partial \tau} = \delta \nabla^2 \dot{v} \tag{8.114}$$

in Ω, with

$$\dot{u}_s = 0, \qquad \dot{v}_s = w(\tau) \tag{8.115}$$

on Ω and

$$\dot{u}(\boldsymbol{\rho}, 0) = \dot{v}(\boldsymbol{\rho}, 0) = 0. \tag{8.116}$$

Then

$$\tfrac{1}{2}(J'(V^0), w) = \int_0^T \{u(\mathbf{0}, \tau) - U(\tau)\}\dot{u}(\mathbf{0}, \tau)\,\mathrm{d}\tau. \tag{8.117}$$

Consider now the adjoint system of equations governing the two adjoint variables $p(\boldsymbol{\rho}, \tau)$ and $q(\boldsymbol{\rho}, \tau)$, namely

$$\frac{\partial p}{\partial \tau} = -\nabla^2 p + \phi^2 R_u p - \{u(\mathbf{0}, \tau) - U(\tau)\}\delta(\mathbf{0}), \tag{8.118}$$

$$\frac{\partial q}{\partial \tau} = -\delta \nabla^2 q + \phi^2 R_v p \tag{8.119}$$

in Ω, with

$$p_s = q_s = 0 \tag{8.120}$$

on $\partial\Omega$ and

$$p(\boldsymbol{\rho}, T) = q(\boldsymbol{\rho}, T) = 0. \tag{8.121}$$

Here $\delta(\mathbf{0})$ is the Dirac measure at the observation point, which we have taken to be the origin, and we note that, as usual, these variables are subject to final rather than initial conditions. Now from eqns (8.117) and (8.118)

$$\begin{aligned}\tfrac{1}{2}(J'(V^0), w) &= \int_0^T \mathrm{d}\tau \iiint_\Omega \mathrm{d}\Upsilon\{u(\mathbf{0}, \tau) - U(\tau)\}\dot{u}(\boldsymbol{\rho}, \tau)\delta(\mathbf{0}) \\ &= \int_0^T \mathrm{d}\tau \iiint_\Omega \mathrm{d}\Upsilon\,\dot{u}(-p_\tau - \nabla^2 p + \phi^2 R_u p).\end{aligned}$$

If we add to the right-hand side of this equation a zero quantity constructed from the integrals of $p \times$ eqn (8.113) $+ q \times$ eqn (8.114) $+ \dot{v} \times$ eqn (8.119), we have

$$\begin{aligned}\tfrac{1}{2}(J'(V^0), w) &= \int_0^T \mathrm{d}\tau \iiint_\Omega \mathrm{d}\Upsilon\{\dot{u}(-p_\tau - \nabla^2 p + \phi^2 R_u p) + \\ &\quad + p(-\dot{u}_\tau + \nabla^2 \dot{u} - \phi^2 R_u \dot{u} - \phi^2 R_v \dot{v}) - \\ &\quad - q(\dot{v}_\tau - \delta\nabla^2 \dot{v}) - \dot{v}(q_\tau + \delta\nabla^2 q - \phi^2 R_v p)\} \\ &= -\delta \int_0^T \mathrm{d}\tau \iint_{\partial\Omega} \mathrm{d}\Sigma \left(\frac{\partial q}{\partial v}\right) w.\end{aligned} \tag{8.122}$$

Thus the necessary condition for optimality is that $V^0(\tau)$ should be so chosen that for every other admissible $V(\tau)$

$$\int_0^T \iint_{\partial\Omega} \left(\frac{\partial q}{\partial v}\right) \{V(\tau) - V^0(\tau)\}\, d\tau\, d\Sigma \leqslant 0. \tag{8.123}$$

8.7. Catalyst decay

Without attempting to survey the whole field of catalyst decay—a subject in which the reader is very well served by Butt's review (1972)—it will be convenient here to note one or two problems of catalyst decay in which the essential transience of the phenomenon is displayed. If there is a slow independent poisoning so that at any instant a predictable distribution of catalytic activity can be calculated, then the steady state results of earlier sections may be invoked (cf. Sections 3.3.10 and 5.5). Such is the case with pore-mouth poisoning where it is assumed that the mouth of the pore is rendered inactive by an adsorbed poison. At any instant there is a dead region on the outside of the catalyst that behaves very much like an additional contribution to the Biot numbers. The isothermal case can be dealt with easily in this way and Ray (1972*b*) has shown how to treat the non-isothermal case. He notes that a partially poisoned pellet may be a more effective catalyst than the unpoisoned one, since the poisoned surface layer keeps the heat within pellet and promotes the reaction rate when the parameters are suitably disposed. However in these cases it is assumed that the poisoning is so slow that the steady state solution can always be used. Such a pseudo-steady state hypothesis has been much used in the theory of gas–solid non-catalytic reactions, and its validity has been examined by Bischoff (1963, 1965*b*) and others.

In type II or type III poisoning (cf. Section 5.1) the poisoning is related to main reaction by a parallel or consecutive reaction and so is present in varying degree throughout the catalytic body. If the effect of the poison is described by a linear blocking of the catalytic sites, every reaction rate is affected by a factor $(1-w)$, where w is the concentration of the poison normalized so that $w = 1$ implies complete poisoning. Let the main reaction be $A_1 \rightarrow A_2$ then, in our standard notation,

$$\frac{\partial u_1}{\partial \tau} = \nabla^2 u_1 - \phi^2(1-w)\{R(u_1, v) + \lambda_1 R_{p1}(u_1, v)\} \tag{8.124}$$

$$\frac{\partial u_2}{\partial \tau} = \delta_2 \nabla^2 u_2 + \phi^2(1-w)\{R(u_1, v) - \lambda_2 R_{p2}(u_2, v)\}, \tag{8.125}$$

$$\frac{\partial v}{\partial \tau} = \nabla^2 v + \beta\phi^2(1-w)R(u_1, v), \tag{8.126}$$

$$\frac{\partial w}{\partial \tau} = \lambda\phi^2(1-w)\{\lambda_1 R_{p1}(u_1, v) + \lambda_2 R_{p2}(u_2, v)\}. \tag{8.127}$$

Here R_{p1} and R_{p2} are the rates of poisoning by the reactions $A_1 \to P$ and $A_2 \to P$ respectively, that is by type II or type III, and they can be taken out of the equations by setting one or other of the λ_i equal to zero. A further constant λ is needed in the last equation, since w was normalized independently. These equations hold in Ω with the usual boundary conditions on $\partial\Omega$ for u_1, u_2, and v, and initial conditions for all the independent variables. If it is assumed that the poisoning is slow (λ small) then it would be sensible to take as initial condition the steady state of the main reaction, since this would be established before any appreciable poisoning had taken place. The effectiveness factor of the main reaction is

$$\eta = \frac{1}{\upsilon} \iiint_{\Omega} (1-w)R(u_1, v)\, \mathrm{d}\Upsilon. \tag{8.128}$$

The possibility of independent fouling by the diffusion and deposition of a poison which is not otherwise related to the main reaction could be included by adding the equation for its concentration

$$\frac{\partial u_3}{\partial \tau} = \delta_3 \nabla^2 u_3 - \phi^2(1-w)\lambda_3 R_{p3}(u_3, v) \tag{8.129}$$

and including a term $\lambda_3 R_{p3}$ in the last equation. It is generally assumed that the heat effect of the poisoning reaction is negligible and that the physical properties are unaffected by the fouling.

Smith and Masamune have calculated some of these profiles for the isothermal sphere (1966) and with Sagara (1967) for the non-isothermal case. In all cases first-order reactions were assumed. Their results may be summarized in Figs. 8.9 and 8.10. In Fig. 8.9 the three mechanisms of parallel ($\lambda_1 \neq 0$, $\lambda_2 = \lambda_3 = 0$), consecutive ($\lambda_2 \neq 0$, $\lambda_3 = \lambda_1 = 0$) and independent ($\lambda_3 \neq 0$, $\lambda_1 = \lambda_2 = 0$) are compared. In the first and third cases the poison builds up on the outside of the sphere and particularly in the last case there is an approximation to pore-mouth poisoning. (The direction of time is shown by the arrows, and the numbers attached to the curves are proportional to the actual values of τ.) In fact Smith and Masamune compared the effectiveness factors as functions of τ with a pore-mouth model, such that the amount of poison laid down was the same in both cases. They found excellent agreement for the independent case, but poor agreement for the consecutive poisoning scheme. Figure 8.10 shows corresponding figures for parallel and consecutive fouling. In the case of parallel fouling

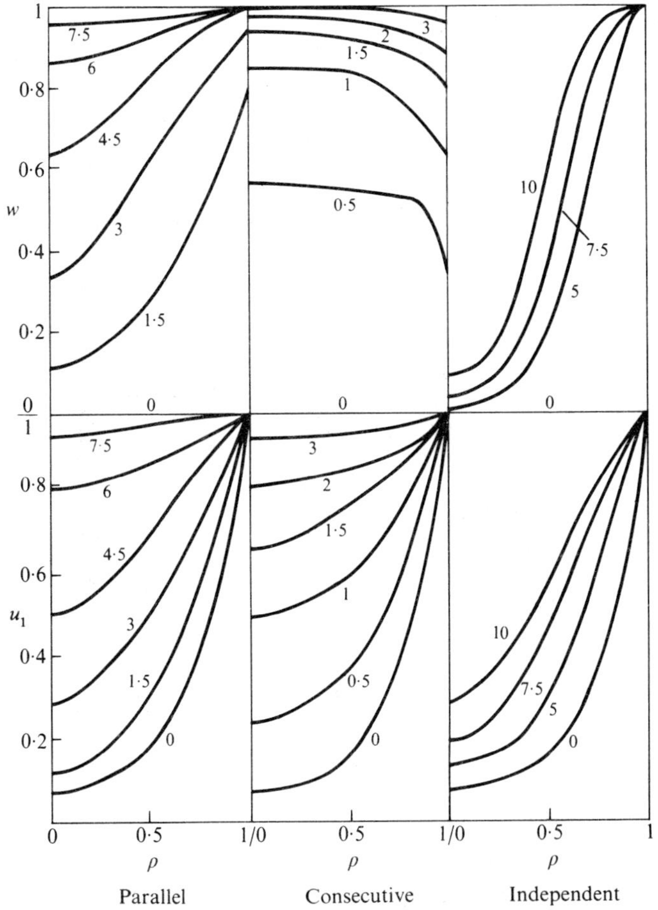

FIG. 8.9. Reactant and poison profiles for parallel, consecutive and independent poisoning by isothermal first-order reactions. (After Smith and Masamune (1966).)

the effect of increasing temperature more than offsets the decrease of reactant concentration and the poison builds up most rapidly in the neighbourhood of $\rho = 0{\cdot}75$. As the poisoning becomes more complete this maximum is less marked. There are of course a great number of parameters in this system, for even if all the reactions are of the first order there is an activation energy associated with each, i.e. $R_{pi} = u_i \exp\{\gamma_i(v-1)/v\}$. Figure 8.10 is drawn for the particular set $\phi = 5$, $\beta = 0{\cdot}1$, $\gamma = 20$, $\gamma_1 = 30$, and it is the fact that $\gamma_1 > \gamma$ which gives the temperature more influence over the poisoning. When the heat of reaction and Thiele modulus are greater the pore-mouth poisoning model is almost exactly fulfilled, whereas with low β and ϕ the build-up of poison is rather uniform throughout the sphere. In the case of

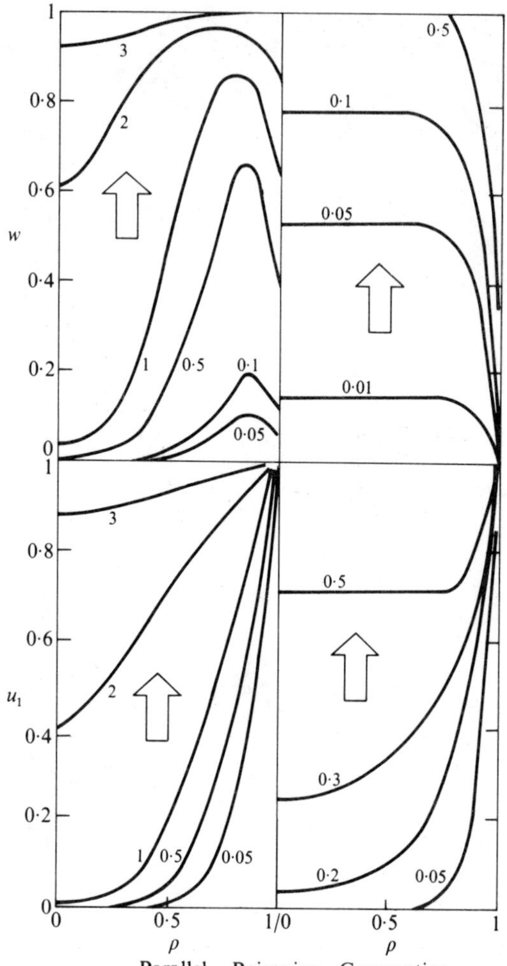

FIG. 8.10. Reactant and poison profiles for parallel and consecutive poisoning by non-isothermal first-order reactions: $\phi = 5$, $\beta = 0{\cdot}1$, $\gamma = 20$, $\gamma_1 = \gamma_2 = 30$. (After Smith, Masamune, and Sagara (1967).)

poisoning by consecutive reactions, it is the centre that first becomes poisoned and as Fig. 8.10 shows, this poisoning is virtually complete. The papers cited contain numerous charts of effectiveness factors as functions of time, and a full discussion of the pore-mouth poisoning approximation.

8.8. The effects of adsorption capacity

Cresswell and Elnashaie (1973) have propounded a lumped model which takes into account the dynamics of the adsorption and surface reaction. It is related to the discrete model considered in Chapter 2 but is even more particular in its attention to the surface concentration. We shall not give the most general presentation with multiple reactions and multicomponent diffusion but consider only the case of a single reactant going irreversibly to products at a rate proportional to its concentration $\hat{c}(\mathbf{r}, t)$ adsorbed on the surface. The concentration $c(\mathbf{r}, t)$ is the concentration of reactant in the free space $\hat{\Omega}$ of Ω (i.e. between $\partial\Omega$, the ostensible external surface, and $\partial\hat{\Omega}$, the reactant surface). This diffuses freely and hence satisfies the equation

$$\nabla \cdot (D\nabla c) = \frac{\partial c}{\partial t} \quad \text{in } \hat{\Omega}, \tag{8.130}$$

with

$$\left(D\frac{\partial c}{\partial n}\right)_{\mathrm{s}} = k_{\mathrm{c}}(c_{\mathrm{f}} - c_{\mathrm{s}}) \quad \text{on } \partial\Omega \tag{8.131}$$

On the internal surface $\partial\hat{\Omega}$ the flux at any point $\mathbf{r}$ is the net rate of adsorption

$$-D\frac{\partial c}{\partial n} = k_{\mathrm{a}}c(\hat{c}_0 - \hat{c}) - k_{\mathrm{d}}\hat{c}, \tag{8.132}$$

where $\hat{c}$ is the surface concentration at complete coverage and the negative sign is needed since the outward normal to $\hat{\Omega}$ points into the solid material. The rate of accumulation of adsorbed reactant is the difference between the rate of adsorption and the rate of reaction

$$\frac{\partial \hat{c}}{\partial t} = k_{\mathrm{a}}c(\hat{c}_0 - \hat{c}) - k_{\mathrm{d}}\hat{c} - \hat{k}\hat{c}, \tag{8.133}$$

where $\hat{k}$ is a first-order reaction rate constant. If the diffusion coefficient is very large, a singular perturbation procedure with respect to a parameter proportional to D^{-1} would be in order. We shall not go through this process in detail for it suffices here to note that the zeroth order term implies that $c(\mathbf{r}, t)$ is constant throughout $\hat{\Omega}$, and that we can work with its mean value. We shall not use any special symbol for this mean value, merely using $c(t)$ to denote its value. From eqns (8.130–2) we have by integrating the first over $\hat{\Omega}$ and using Green's formula

$$\begin{aligned} V_{\mathrm{p}}\varepsilon\frac{\mathrm{d}c}{\mathrm{d}t} &= \iiint\limits_{\hat{\Omega}} \nabla \cdot (D\nabla c)\,\mathrm{d}V = \iint\limits_{\partial\Omega} D\frac{\partial c}{\partial n}\,\mathrm{d}S + \iint\limits_{\partial\hat{\Omega}} D\frac{\partial c}{\partial n}\,\mathrm{d}S \\ &= k_{\mathrm{c}}S_x(c_{\mathrm{f}} - c) - \rho_{\mathrm{b}}S_{\mathrm{g}}V_{\mathrm{p}}\{k_{\mathrm{a}}c(\hat{c}_0 - \hat{c}) - k_{\mathrm{d}}\hat{c}\}. \end{aligned} \tag{8.134}$$

The suffix s may be dropped from c_{s} in eqn (8.131) since it also equals the mean concentration. In the heat balance we will assume from the start that the conductivity is large so that the temperature is $T(t)$ throughout $\hat{\Omega}$ and on $\partial\hat{\Omega}$ and $\partial\Omega$. Then

$$V_p C_p \frac{dT}{dt} = hS_x(T_f - T) + (-\Delta H)_a \rho_b S_g V_p \{k_a c(\hat{c}_0 - \hat{c}) - k_d \hat{c}\} + \\ + (-\Delta H)_r \rho_b S_g V_p \hat{k}\hat{c}, \tag{8.135}$$

where $(\Delta H)_r$ and $(\Delta H)_a$ are the heats of surface reaction and adsorption and C_p the heat capacity per unit volume. The heat of reaction as normally given will include both $(\Delta H)_r$ and $(\Delta H)_a$ so we may write

$$\Delta H = (\Delta H)_r + (\Delta H)_a. \tag{8.136}$$

To render the equations dimensionless let $\hat{k}_f$ denote the value of $\hat{k}$ at $T = T_f$ and set

$$u = \frac{c}{c_f}, \quad \hat{u} = \frac{\hat{c}}{\hat{c}_0}, \quad v = \frac{T}{T_f}, \quad \tau' = \frac{\rho_b S_g \hat{k}_f \hat{c}_0}{c_f} t, \\ \beta = \frac{(-\Delta H)\hat{c}_0}{C_p T_f}, \quad \hat{\beta} = \frac{(-\Delta H)_a \hat{c}_0}{C_p T_f}, \quad \gamma = \frac{E}{RT_f}, \quad \hat{\varepsilon} = \frac{\rho_b S_g \hat{c}_0}{c_f}, \tag{8.137}$$

$$\hat{\kappa} = \frac{\hat{k}}{\hat{k}_f}, \quad \kappa = \frac{k_d}{k_a c_f}, \quad \lambda = \frac{k_a c_f}{\hat{k}_f}, \\ \mu' = \frac{hS_x c_f}{\hat{k}_f \rho_b S_g V_p C_p \hat{c}_0}, \quad v' = \frac{k_c S_x c_f}{\hat{k}_f \rho_b S_g V_p \hat{c}_0}. \tag{8.138}$$

The parameter $\hat{\varepsilon}$ is a measure of the adsorption capacity. Then

$$\hat{\varepsilon}\frac{d\hat{u}}{d\tau'} = \lambda\{u(1-\hat{u}) - \kappa\hat{u}\} - \hat{\kappa}\hat{u}, \tag{8.139}$$

$$\varepsilon\frac{du}{d\tau'} = v'(1-u) - \lambda\{u(1-\hat{u}) - \kappa\hat{u}\}, \tag{8.140}$$

$$\frac{dv}{d\tau'} = \mu'(1-v) + \hat{\beta}\lambda\{u(1-\hat{u}) - \kappa\hat{u}\} + (\beta - \hat{\beta})\hat{\kappa}\hat{u}. \tag{8.141}$$

These equations are conveniently written

$$\frac{d}{d\tau'}(\varepsilon u + \hat{\varepsilon}\hat{u}) = v'(1-u) - \hat{\kappa}\hat{u}, \tag{8.142}$$

$$\frac{d}{d\tau'}(v + \hat{\beta}\hat{\varepsilon}\hat{u}) = \mu'(1-v) + \beta\hat{\kappa}\hat{u}, \tag{8.143}$$

for then if the assumption is made that adsorption equilibrium is very rapid (i.e. $\lambda \to \infty$), the third equation can be replaced by the adsorption equilibrium relationship

$$\hat{u} = \frac{u}{u+\kappa}. \tag{8.144}$$

Furthermore if $\kappa \gg 1$ this can be regarded as approximately linear $\hat{u} = u/\kappa$. Even if we take the linear limit and write

$$(\varepsilon + \hat{\varepsilon}/\kappa)\frac{\mathrm{d}u}{\mathrm{d}\tau'} = \nu'(1-u) - (\hat{\kappa}/\kappa)u$$

for the mass balance, we would still have to make the further assumption $\hat{\beta} = 0$ to get

$$\frac{\mathrm{d}v}{\mathrm{d}\tau'} = \mu'(1-v) + \beta(\hat{\kappa}/\kappa)u$$

and make the pair of equations look like those of a stirred tank with a first-order reaction, namely

$$\begin{aligned} \frac{\mathrm{d}u}{\mathrm{d}\tau} &= 1 - u - \psi^2 \hat{\kappa} u, \\ L\frac{\mathrm{d}v}{\mathrm{d}\tau} &= 1 - v + \beta'\psi^2 \hat{\kappa} u, \\ \hat{\kappa} &= \exp\{\gamma(v-1)/v\}, \end{aligned} \tag{8.145}$$

where

$$\tau = \nu'\tau'/(\varepsilon + \hat{\varepsilon}/\kappa), \quad L = \nu'/\mu'(\varepsilon + \hat{\varepsilon}/u), \qquad \psi^2 = 1/\kappa\nu', \quad \beta' = \nu'\beta/\mu'.$$

If we do not go to the linear limit but use the Langmuir isotherm (8.144), we have to assume $\hat{\varepsilon} \ll \varepsilon$ to get the equations for a stirred tank with Langmuir–Hinshelwood kinetics. In this case we might take

$$\begin{gathered} \tau = \nu'\tau'/\varepsilon, \quad L = \nu'/\mu'\varepsilon, \quad \psi^2 = 1/\nu', \quad \beta' = \nu'\beta/\mu', \\ R(u, v) = \hat{\kappa}u/(u+\kappa) \end{gathered} \tag{8.146}$$

to give

$$\begin{aligned} \frac{\mathrm{d}u}{\mathrm{d}\tau} &= 1 - u - \psi^2 R(u, v), \\ L\frac{\mathrm{d}v}{\mathrm{d}\tau} &= 1 - v + \beta'\psi^2 R(u, v). \end{aligned} \tag{8.147}$$

Let us recall that the conditions for the stability of a steady state are

$$\beta'\psi^2 R_v < \min\{1 + \psi^2 R_u, 1 + L(1 + \psi^2 R_u)\}, \tag{8.148}$$

so that a steady state which is stable for $L = 1$ may become unstable if

$$\beta'\psi^2 R_v > 1, \qquad L < L_c = (\beta'\psi^2 R_v - 1)/(1+\psi^2 R_u). \tag{8.149}$$

Compare now the situation when the assumption $\hat{\varepsilon} \ll \varepsilon$ is not made. Using the relation (8.144) and the variables (8.146) we have

$$\frac{du}{d\tau}\left\{1+\frac{\hat{\varepsilon}}{\varepsilon}\frac{\kappa}{(\kappa+u)^2}\right\} = 1-u-\psi^2 R(u,v), \tag{8.150}$$

$$L\frac{dv}{d\tau}+\frac{L\hat{\varepsilon}\hat{\beta}\kappa}{(\kappa+u)^2}\frac{du}{d} = 1-v+\beta'\psi^2 R(u,v). \tag{8.151}$$

We observe that the concentration and temperature at the steady state, u_e and v_e, are unchanged by the terms in $\hat{\varepsilon}$. Letting

$$u^* = u-u_e, \qquad v^* = v-v_e, \tag{8.152}$$

the linearized equations for u^* and v^* are

$$\left\{1+\frac{\hat{\varepsilon}}{\varepsilon}\frac{\kappa}{(\kappa+u_e)^2}\right\}\frac{du^*}{d\tau} = -(1+\psi^2 R_u)u^* - \psi^2 R_v v^*, \tag{8.153}$$

$$\begin{aligned}\frac{dv^*}{d\tau} = &\left\{\frac{\beta'\psi^2}{L}R_u+\frac{\hat{\varepsilon}\hat{\beta}\kappa}{(\kappa+u_e)^2}(1-\psi^2 R_u)\right\}u^* - \\ &-\left\{\frac{1}{L}-\frac{\beta'\psi^2}{L}R_v-\frac{\hat{\varepsilon}\hat{\beta}\kappa}{(\kappa+u_e)^2}\psi^2 R_v\right\}v^*.\end{aligned} \tag{8.154}$$

The stability condition

$$\beta'\psi^2 R_v < 1+\psi^2 R_u$$

is unchanged by the greater complication of this equation, but the condition $\beta'\psi^2 R_v < 1+L(1+\psi^2 R_u)$ becomes

$$\begin{aligned}\beta'\psi^2 R_v &< 1+L(1+\psi^2 R_u)\left\{1+\frac{\hat{\varepsilon}}{\varepsilon}\frac{\kappa}{(\kappa+u_e)^2}\right\}^{-1} - \frac{\hat{\varepsilon}\hat{\beta}\kappa}{(\kappa+u_e)^2}\psi^2 R_v \\ &= 1+L(1+\psi^2 R_u) - \frac{\hat{\varepsilon}\kappa}{\varepsilon(\kappa+u_e)^2}\Bigg[\varepsilon\hat{\beta}\psi^2 R_v + L(1+\psi^2 R_u)\times \\ &\quad\times\left\{1+\frac{\hat{\varepsilon}}{\varepsilon}\frac{\kappa}{(\kappa+u_e)^2}\right\}^{-1}\Bigg].\end{aligned} \tag{8.155}$$

Since all the terms in the square bracket are positive, the effects of both $\hat{\varepsilon}$, the adsorption capacity, and $\hat{\beta}$, the heat of adsorption, are to make it more easy for the system to become unstable. If $\hat{L}_c$ is the critical 'Lewis number' with these effects considered, and L_c, given by eqn (8.149), is the conventional value,

$$\hat{L}_c = L_c(1+E)\left\{1+E\frac{\hat{\beta}\varepsilon\psi^2 R_v}{\beta'\psi^2 R_v - 1}\right\} > L_c(1+E)\left\{1+E\frac{\hat{\beta}\varepsilon R_v}{R_u}\right\}, \qquad (8.156)$$

where

$$E = \hat{\varepsilon}\kappa/\varepsilon(\kappa+u_e)^2.$$

Cresswell and Elnashaie give some examples of the changed trajectories in the phase plane produced by these effects.

8.9. Extremely rapid variations of temperature

If we focus attention on what is believed to happen at the site of a heterogeneously catalysed reaction we have to recognize the possibility of extremely rapid, and perhaps large, variations of temperature. These are of interest both practically and theoretically: practically, because high temperatures may account for the sintering of catalysts or the clumping together of crystallites; theoretically, because the extreme rapidity of the changes requires the use of the telegrapher's equation rather than the usual diffusion equation. In supported metal catalysts, such as platinum dispersed in alumina, the sites of reaction are small crystals, perhaps roughly spherical in shape, 10–100 Å, and attached to the non-metallic support. When a molecule reacts at such a site, heat is released during a very short interval of time and has to be dissipated through the crystal and the support. Luss (1970) speaks of a highly exothermic reaction ($-\Delta H = 60\,000$ cal g mol^{-1}) which might release 10^{-22} cal s^{-1} at each site, but would do so in a time interval of the order of 10^{-13} s. This very rapid release of heat would be followed by a relatively long period, of the order of 10^{-3} s, before the next reaction occurs. The problem is to estimate the maximum temperature that can be reached in the crystallite to see if this can account for the sintering effects. Luss suggested two models which though very simplified might serve to bracket the effect. In the first he considered the crystallite to be a small disc with one face attached to the support at constant temperature and the other subjected to pulses of heat of duration t_r at intervals of the t_p. In the second he considered the heat pulse to be applied to a small circular patch in the plane and the heat to be dissipated by conduction away from it in the plane. The first model would be likely to overestimate the temperature rise and the second to underestimate it.

Luss uses the normal diffusion equation derived from Fourier's Law of heat conduction

$$\frac{\partial T}{\partial t} = \alpha\nabla^2 T, \qquad (8.157)$$

where α is the thermal diffusivity, but Chan, Low, and Mueller (1971) and Ruckenstein and Petty (1972) suggest that the changes are so rapid that the finite propagation speed of the heat should be taken into account. In this case it is not the heat flux $\mathbf{j}$ which is proportional to the gradient, but $\mathbf{j}+\tau_h\partial\mathbf{j}/\partial t$, where τ_h is the relaxation time often written as α/c^2, and c is a propagation speed. Thus eqn (8.157) gives place to

$$\frac{1}{c^2}\frac{\partial^2 T}{\partial t^2}+\frac{1}{\alpha}\frac{\partial T}{\partial t} = \nabla^2 T \tag{8.158}$$

and the two are only equivalent in the limit $c \to \infty$. Because the propagation speed is finite, the heat cannot be dissipated so rapidly and the temperature rise is greater. It will suffice to give the solution to the telegrapher's equation since the corresponding solution to eqn (8.157) should be deducible from the limit as $c \to \infty$.

Chan, Low, and Mueller (1971) have solved the problem for Luss's first model in which

$$\frac{1}{c^2}\frac{\partial^2 T}{\partial t^2}+\frac{1}{\alpha}\frac{\partial T}{\partial t} = \frac{\partial^2 T}{\partial x^2}, \quad 0 < x < L, \tag{8.159}$$

with

$$T(x,0) = \frac{\partial T}{\partial t}(x,0) = 0, \quad 0 < x < L, \tag{8.160}$$

$$T(0,t) = 0,$$

and

$$k\frac{\partial T}{\partial x} = f(t), \qquad x = L \tag{8.161}$$

The time-varying gradient at the boundary arises from generating an amount of heat q per unit area for a duration of t_r at intervals of t_p. The flux at $x = L$ is thus

$$f(t) = \begin{cases} -q/t_r & (rt_p < t < rt_p+t_r), \\ 0 & (rt_p+t_r < t < (r+1)t_p), \end{cases} \tag{8.162}$$

for $r = 0, 1, 2, \ldots$. Note however that with the finite propagation speed it is not j that is equal to $-k(\partial T/\partial x)$, but $j+(\alpha/c^2)(\partial j/\partial t)$. Thus

$$-f(t) = \left(1+\frac{\alpha}{c^2}\frac{\partial}{\partial t}\right)j, \tag{8.163}$$

where k is the conductivity and f the discontinuous function given by eqn (8.162).

These equations can be solved by the use of the Laplace transform to give

$$T(x,t) = \sum_{n=0}^{\infty} \frac{4\alpha(-1)^n}{kLM_n} \sin(2n+1)\frac{\pi x}{2L} \int_0^t f(t-t') \exp\left(\frac{c^2t'}{2\alpha}\right) \sinh\left(M_n \frac{c^2t'}{2\alpha}\right) dt' \tag{8.164}$$

where

$$M_n^2 = 1 - \{(2n+1)\pi\alpha/cL\}^2.$$

We expect the greatest temperature to be when the pulse has just ceased, $t = rt_p + t_r$, on the surface where the heat is generated, $x = L$, and after the steady oscillations have been built up, $r \to \infty$. In this limit and with $t_p \gg t_r$, $c^2t_p(1 \pm M_n) \gg 2\alpha$

$$\frac{kT_{\max}}{L(q/t_r)} = 1 - \frac{8}{\pi^2} \sum_{n=0}^{\infty} \frac{1}{(2n+1)^2} \exp\{(-c^2t_r)/2\alpha\} \times$$
$$\times \left[\cosh M_n \frac{c^2t_r}{2\alpha} + \left(2M_n - \frac{1}{M_n}\right) \sinh M_n \frac{c^2t_r}{2\alpha}\right] \tag{8.165}$$

The limit of this equation when $c \to \infty$ is obtained by noting that

$$(1-M_n)\frac{c^2t_r}{2\alpha} \to \frac{(2n+1)^2\pi^2\alpha t_r}{4L^2} = m_n \frac{\alpha t_r}{L^2},$$

and hence that

$$\frac{kT_{\max}}{L(q/t_r)} = 1 - \frac{8}{\pi^2} \sum_{n=0}^{\infty} \frac{1}{(2n+1)^2} \exp\{(-m_n\alpha t_r)/L^2\}. \tag{8.166}$$

This agrees with Luss's formula when $m_n\alpha t_p/L^2 \gg 1$.

Clearly three dimensionless quantities are involved here:

$V = kT_{\max}/qc$ = dimensionless temperature rise,

$\tau_r = c^2t_r/2\alpha$ = dimensionless reaction time,

$\Lambda = cL/2\alpha$ = dimensionless thickness.

We note that τ_r/Λ is the reaction time as a fraction of the propagation time and that when $c \to \infty$ the dimensionless reaction time is $\alpha t_r/L^2 = \tau_t/2\Lambda^2$. Then

$$V = \frac{\Lambda}{\tau_r}\left\{1 - \frac{8}{\pi^2} \sum_{n=0}^{\infty} \frac{1}{(2n+1)^2} \exp(-\tau_r) \left[\cosh M_n\tau_r + \left(2M_n - \frac{1}{M_n}\right) \sinh M_n\tau_r\right]\right\}, \tag{8.167}$$

where

$$M_n = \left\{1 - \frac{(2n+1)^2\pi^2}{4\Lambda^2}\right\}.$$

When τ_r is fairly large the parabolic equation gives practically the same result as the hyperbolic. For example, when $\tau_r = 15$ and $\Lambda > 5$ the temperature rise is fairly constant at $V = 0{\cdot}206$. This corresponds to the limit given by Luss of

$$V \sim \left(\frac{2}{\pi\tau_r}\right)^{\frac{1}{2}}, \quad \Lambda \text{ large.} \tag{8.168}$$

But when τ_r and Λ are small, V can be very much larger. For example, $\tau_r = 0{\cdot}05$, $V = 11$.

Luss quotes typical values for the physical constants of $k = 0{\cdot}175$ cal s^{-1} cm^{-1} $°C^{-1}$, $\alpha = 0{\cdot}254$ cm^2 s^{-1}, $q = 5 \times 10^{-5}$ cal cm^{-2}, $t_r = 10^{-13}$ s, $t_p = 10^{-3}$ s. He finds maximum temperature rises of 285, 475, and 511°C for $L = 10$, 20, and 30 Å respectively. Chan, Low, and Mueller (1971) suggest two values for the propagation velocity c. The first is the velocity of a free electron gas at moderate temperature, $8{\cdot}7 \times 10^6$ cm s^{-1}, which is probably an upper bound. The second is 5×10^5 cm^{-1}, the speed of sound in crystals, which is probably a lower bound. With the larger value of c they obtained figures close to those of Luss, but with the smaller value the temperature rise with $t_r = 10^{-13}$, $L = 30$ Å was calculated to be two or three times greater than that given by the parabolic equation. Luss's lower bound for the circular path of 18 Å in radius was 49°C.

Ruckenstein and Petty (1972) assumed the crystallite to be a small sphere whose surface was heated during the reaction period t_r. They relate the average temperature of the sphere to the heat generated in the pulse and then note that if the speed of propagation is finite, the heat will be confined to an outer annulus of thickness ct_r, and that the mean temperature here will be accordingly greater. In addition the number of available sites on the surface of the sphere may increase as its temperature increases and this can greatly augment the effect.

More interesting is their solution of the cooling of the surface of a sphere according to the telegrapher's equation. If the sphere is initially at a constant temperature T_0 and its surface is maintained at a lower temperature T_s we may take

$$V = (T_0 - T)/(T_0 - T_s), \quad \rho = r/a, \quad \tau = ct/a, \quad \gamma = ca/2\alpha \tag{8.169}$$

Then

$$\frac{\partial^2 v}{\partial \tau^2} + 2\gamma \frac{\partial v}{\partial \tau} = \frac{1}{\rho^2}\frac{\partial}{\partial \rho}\left(\rho^2 \frac{\partial v}{\partial \rho}\right), \tag{8.170}$$

with

$$v(\rho, 0) = v_\tau(\rho, 0) = 0, \qquad v(1, \tau) = 1, \qquad v(0, \tau) < \infty. \tag{8.171}$$

This may be solved by the Laplace transform

$$\bar{v}(\rho, \sigma) = \int_0^\infty e^{-\sigma\tau} v(\rho, \tau)\, d\tau, \tag{8.172}$$

giving

$$\bar{v}(\rho, \sigma) = \frac{1}{\sigma} \frac{\sinh \lambda(\sigma)\rho}{\rho \sinh \lambda(\sigma)}, \tag{8.173}$$

where

$$\lambda^2(\sigma) = \sigma^2 + 2\gamma\sigma. \tag{8.174}$$

In particular the centre temperature has a transform of

$$\bar{v}(0, \sigma) = \frac{\lambda(\sigma)}{\sigma} \operatorname{cosech} \lambda(\sigma) = 2 \sum_{m=1}^{\infty} \frac{\lambda(\sigma)}{\sigma} e^{-(2n+1)\lambda(\sigma)} \quad .$$

and

$$\begin{aligned} v(0, \tau) = 4\gamma \sum_{n=0}^{\infty} e^{-\gamma\tau} I_0[\gamma\surd\{\tau^2 - (2n+1)^2\}] H(\tau - 2n - 1) + \\ + 2 \sum_{n=0}^{\infty} \frac{d}{d\tau}\left(e^{-\gamma\tau} I_0[\gamma\surd\{\tau^2 - (2n+1)^2\}] H(\tau - 2n - 1)\right), \end{aligned} \tag{8.175}$$

and H is the Heaviside step function. This formula shows that $v(0, \tau)$ is unchanged until $\tau = 1$ due to the finite speed of propagation. The finiteness of the speed of propagation again enhances the temperature and prolongs its influence in the outer shell of the sphere.

8.10. Diffusion waves

The presence of an autocatalytic step in a reaction scheme can often lead to wave-like solutions of the diffusion equation. The simplest example arose from a problem in the genetical theory of natural selection and was first treated by no lesser mathematicians than Kolmogoroff, Petrovsky, and Piscounoff (1937) who published one of the fundamental papers in the theory of intermediate asymptotics. Zel'dovich and Frank-Kamenetskii (1938) were quick to see the application of these ideas to the theory of combustion and flame propagation and similar methods have been applied to other problems of mathematical physics as the review of Zel'dovich and Barenblatt (1971)

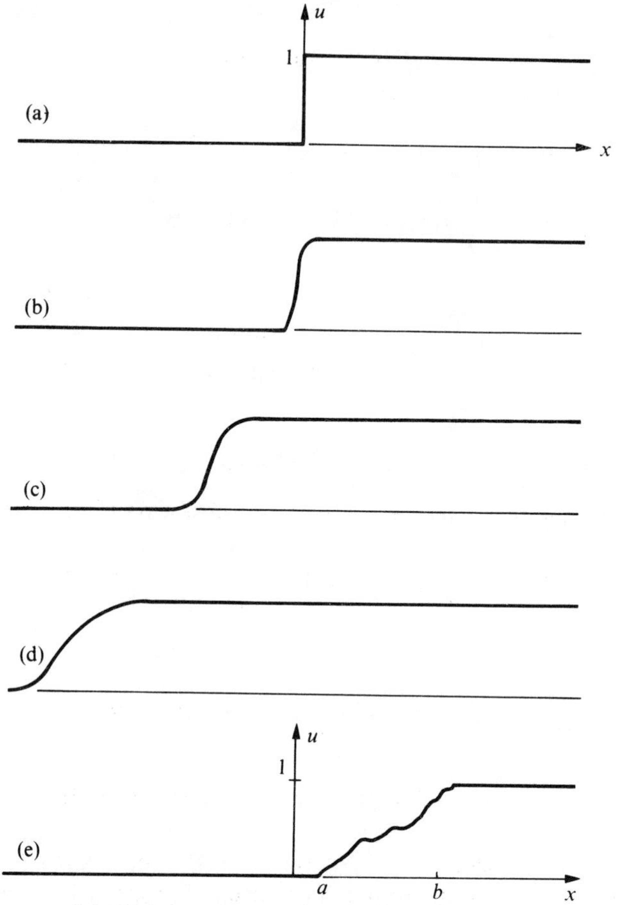

FIG. 8.11. Progress of the Kolmogoroff–Petrovsky–Piscounoff diffusion wave: (a) initial condition, (b) inception, (c) development, (d) asymptotic form, (e) arbitrary initial conditions.

shows. The problem has considerable literature to which some reference is made at the end of the chapter. A similar but kinetically more complicated oxidation of malonic acid, the so-called Belousov–Zhabotinskii reaction, has recently received a considerable amount of attention. There has also been a large amount of work done by the Brussels school on dissipative structures and the breakdown of symmetry. We shall propound some of the ideas of these three topics in an introductory way—to do them proper justice would require a monograph in itself.

The basic idea of the Kolmogoroff–Petrovsky–Piscounoff diffusion wave can be seen in Fig. 8.11. Let $u(x, t)$ be the concentration of a substance produced by an irreversible autocatalytic reaction. It is initially confined to the right half of an infinite column and the left half is filled with a stoicheiometric concentration of the reactant from which it is made. At time $t = 0$ a barrier is removed and the product starts to diffuse into the reactant. But the state of pure reactant is an unstable equilibrium state and the smallest amount of product will catalyse the reaction which, being irreversible, proceeds until it reaches the stable equilibrium state of pure product. This sets up a concentration wave which moves to the left, and we shall show that it ultimately has constant shape and speed. There is a lower bound to this speed, and waves of this speed are stable in the sense that they are the asymptotic form of the solution, when the initial conditions are as shown in the lowest part of the figure with $u(x, 0) = 0, x \leqslant a, u(x, 0) = 1, x > b$.

Let $c(x, t)$ be the concentration of product and c_{f} the total concentration of reactant and product, which we assume to be constant everywhere. This concentration satisfies

$$\frac{\partial c}{\partial t} = D\frac{\partial^2 c}{\partial x^2} + r(c), \tag{8.176}$$

where the rate of formation of the product $r(c)$ will be zero when $c = 0$, since the reaction is autocatalytic, and when $c = c_{\mathrm{f}}$, since it is irreversible. For convenience we may define r_{m} to be the maximum rate of reaction achieved at some intermediate concentration and set

$$\xi = x\left\{\frac{r_{\mathrm{m}}}{Dc_{\mathrm{f}}}\right\}^{\frac{1}{2}}, \quad \tau = t\frac{r_{\mathrm{m}}}{c_{\mathrm{f}}}, \quad u = \frac{c}{c_{\mathrm{f}}}, \quad R(u) = \frac{r(c)}{r_{\mathrm{m}}}, \tag{8.177}$$

then

$$\frac{\partial u}{\partial \tau} = \frac{\partial^2 u}{\partial \xi^2} + R(u), \tag{8.178}$$

where

$$R(0) = R(1) = 0, \quad \text{and} \quad 0 < R(u) \leqslant 1, \quad 0 < u < 1. \tag{8.179}$$

In addition we shall assume that $R(u)$ is sufficiently continuous and that

$$R'(u) < R'(0) = \alpha^2. \tag{8.180}$$

These last two equations imply that $R'(1) < 0$.

Let us look for a solution of the form

$$u(\xi, \tau) = U(\xi + \kappa\tau + \mu), \tag{8.181}$$

(where κ and μ are constants) which has the limiting values of zero and 1 as $\xi \to -\infty$ and $+\infty$ respectively. Clearly μ can be arbitrary, but we expect κ to emerge from the analysis. Denoting the argument of U by ζ, we see that the function $U(\zeta)$ must satisfy

$$U'' - \kappa U' + R(U) = 0, \tag{8.182}$$

with $U \to 0$ as $\zeta \to -\infty$, $U \to 1$ as $\zeta \to \infty$.

The equation may be written as a pair of equations

$$\begin{aligned} U' &= V, \\ V' &= \kappa V - R(U), \end{aligned} \tag{8.183}$$

and its solution represented in the phase plane of U and V. In fact, since $U' \to 0$ as $\xi \to \pm\infty$, we want a solution joining the origin in Fig. 8.12 with the point C, (1, 0). The equation may be written

$$\frac{dV}{dU} = \kappa - \frac{R(U)}{V}, \tag{8.184}$$

and the isoclines in the (U, V)-plane will be given by different scalings of the curve $R(U)$. For if $(dV/dU) = s$, the point (U, V) must lie on $V = R(U)/(\kappa - s)$; in particular the locus $s = 0$ is the broken line shown and the loci for $s = \kappa$ and $s = \infty$ are two verticals $U(1-U) = 0$ and the axis $V = 0$ respectively.

We note first that C is a saddle-point, for the linearization about it gives

$$(U-1)' = V, \quad V' = R'(1)(U-1) + \kappa V,$$

so that the eigenvalues are roots of

$$\lambda^2 + \kappa\lambda + R'(1) = 0$$

The positive root of this equation is

$$\lambda = \frac{\kappa}{2} + \left\{ \left(\frac{\kappa}{2}\right)^2 - R'(1) \right\}^{\frac{1}{2}}$$

so that there is a trajectory coming into C tangentially to the line $V = \lambda(1-U)$. We also observe that $\lambda < -R'(1)/\kappa$ so that this trajectory comes in beneath the broken curve in Fig. 8.12. By contrast the eigenvalues of the linearization at O,

$$U' = V, \qquad V' = \alpha^2 U + \kappa V,$$

are both negative, being roots of

$$\lambda^2 + \kappa\lambda + \alpha^2 = 0,$$

and the origin is an unstable node if

$$\kappa > 2\alpha. \tag{8.185}$$

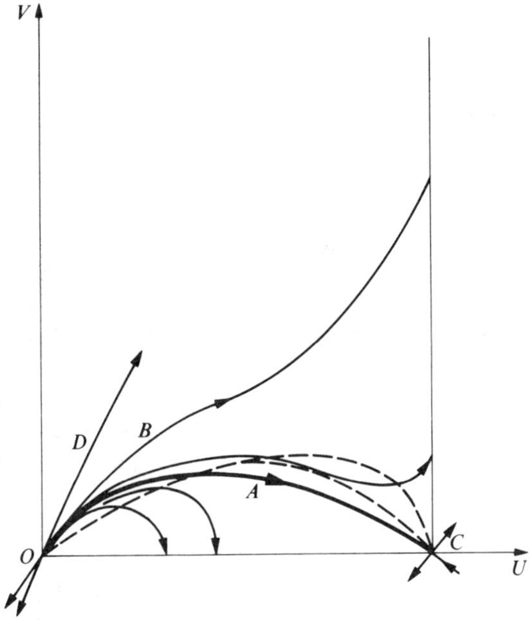

FIG. 8.12. The phase plane of U and V for the Kolmogoroff–Petrovsky–Piscounoff diffusion wave. OAC corresponds to the asymptotic waveform. Locus of maxima and minima of trajectories shown as broken line.

We also note that the two directions of flight from the origin (the lines OD and OB) are $V = \lambda U$, and that both values of λ are greater than $(2\alpha/\kappa)$, the slope of the locus $s = 0$ (the broken curve in Fig. 8.12). The disposition of the trajectories in the region of interest, $V \geqslant 0, 0 \leqslant U \leqslant 1$, must therefore be as shown in the figure, and in particular there will be one, OAC, which passes from O to C and gives the required solution. If $\kappa = 2\alpha$ the two eigenvalues at O coincide but the picture is not substantially changed. It follows that there is a solution of constant form (8.181) to eqn (8.178) for any wave velocity $\kappa \geqslant 2\alpha$. The actual form of the diffusion wave has to be computed by a further integration. Denoting by $\zeta_{\frac{1}{2}}$ the point at which $U(\zeta) = \frac{1}{2}$, we have

$$\zeta = \zeta_{\frac{1}{2}} + \int_{\frac{1}{2}}^{U(\zeta)} \frac{\mathrm{d}U}{V(U)}. \tag{8.186}$$

We note also that the larger the value of κ the less will be the slope of OB and the lower the broken curve of zero slope. It follows that $V(U)$ will decrease for any given U as κ increases and the faster moving waves will have a more gradual slope.

The complete proof that the wave form with the lowest velocity $\kappa = 2\alpha$ is stable is lengthy and will not be given in detail. It is stability in the sense that this is the form which is approached by the solution of eqn (8.178) with initial values that are zero to the left of a finite interval and 1 to the right, as shown in Fig. 8.11(e). After certain existence, uniqueness, and monotonicity theorems, Kolmogoroff, Petrovsky, and Piscounoff consider the function

$$u^*(\xi, \tau) = u(\xi + \chi(\tau), \tau),$$

where $\chi(\tau)$ is chosen so as to keep the value u at $\xi = -\chi(\tau)$ constant. It is then shown that $u^*(\xi, \tau)$ tends uniformly to a function $u^*(\xi)$ of ξ as $\tau \to \infty$. The derivative of $\chi(\tau)$ is shown to approach 2α and the form of u^* to be that of U when $\kappa = 2\alpha$. Kanel (1960) has proved a similar theorem for a more general form of $R(u)$ which can be negative in $0 < u < u_* < 1$, but has a positive integral over the complete interval from 0 to 1.

The Belousov reaction as modified by Zhabotinskii is much more complicated than the simple autocatalytic reaction used in the previous example. Its mechanism and kinetics are not completely understood but a simplified model that satisfactorily saves the phenomena has been devised by Zhabotinskii, Zaikin, Korzukhin, and Kreitser (1971). The oscillations, which are easily observable by colour changes and measureable by light absorption in the near ultra-violet, are in the concentrations of two forms of cerium ion, X and X′, and two species, Y and Z, which respectively promote and inhibit the conversion of X′ to X. At least two other species are present: A is involved in the oxidation of X′ to X which is catalysed by one of its products Y; B is

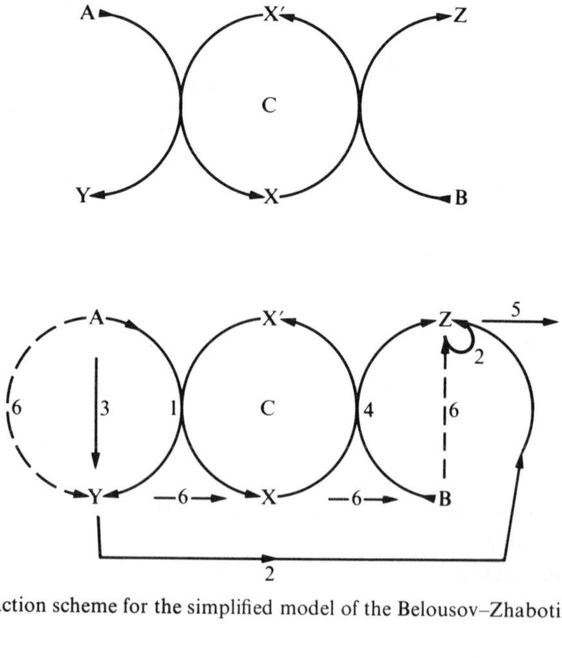

FIG. 8.13. Reaction scheme for the simplified model of the Belousov–Zhabotinskii reaction.

involved in the reduction of X to X′, which liberates Z the inhibitor of oxidation. The basic kinetic scheme is illustrated in Fig. 8.13 at the top. In the laboratory A is a bromate, B bromomalonic acid, C is the total cerium $X \cup X'$, X′ is Ce^{3+}, X is Ce^{4+} and Z may be Br^-. Using lower case letters for the concentrations, a, b, and $c = x'+x$ are to be regarded as constants, and x, y, and z the dependent variables in a set of ordinary differential equations representing the kinetics. The autocatalytic effect of Y is incorporated by writing the rate of formation of X from X′ as proportional to $x'y$ as well as to a and the inhibitory effect of Z is obtained by assuming that Y and Z combine to form an inactive species. The active species is supposed to arise from A so that the rate of change of y is given by the sum of these three steps as

$$\dot{y} = k_1 ax'y - k_2 yz + k_3 a. \tag{8.187}$$

The formation of X from X′ is at the rate $k_1 ax'y = k_1 a(c-x)y$ and the reduction as proportional to the concentrations of B and X; thus

$$\dot{x} = k_1 a(c-x)y - k_4 bx. \tag{8.188}$$

The species Z is formed by this last reaction but disappears spontaneously; its deactivation is not however counted as a disappearance. Z is also formed by a series of background reactions whose rate is proportional to a function of x, y, a, and b. This last term is unabashedly empirical but gives for the rate of formation of Z,

$$\dot{z} = k_4 bx - k_5 z + k_6 ax(y-k)^2, \tag{8.189}$$

where k is a function of a and b. The full kinetic scheme is shown in Fig. 8.13, at the bottom, with this last step as the broken line joining A, Y, X, B and going to Z. After making a pseudo-steady state assumption on the concentration of Z, Zhabotinskii makes certain assumptions on the magnitudes and relationships of the ks and arrives at two equations for the dimensionless variables $u = x/c$ and $v = k_6 y$, namely

$$\begin{aligned} \dot{u} &= \beta v(1-u) - \gamma u, \\ \varepsilon \dot{v} &= \beta v[1 - u\{1+\alpha+(v-\alpha)^2\}] + \varepsilon\beta, \end{aligned} \tag{8.190}$$

where α, β, γ, and ε are constants depending on the k's and a, b, and c. The oscillations given by the eqns (8.190) agreed quite well with observation over a wide range of parameter space so that, in spite of its empirical element, it is worthy of a considerable degree of confidence.

Whilst good agreement has been obtained in the batch, the time-dependent diffusion equations which would be analogous to these have not yet been analysed. However certain simplified forms have been studied by Romanovskii and Sidarova (1966) who took the Lotka scheme $A \to Y \to X \to B$, in which the second step is autocatalytic, and formed the equations

$$\frac{\partial x}{\partial t} = D_1\nabla^2 x + k_1 xy - k_2 x,$$
$$\frac{\partial y}{\partial t} = D_2\nabla^2 y - k_1 xy + k_0. \tag{8.191}$$

There is a uniform steady state for these equations with $x_e = k_0/k_2$, $y_e = k_2/k_1$. Setting

$$x = x_e(1+u), \qquad y = y_e(1+v),$$
$$\boldsymbol{\rho} = \mathbf{r}(k_2/D_1)^{\frac{1}{2}}, \quad \tau = k_2 t, \quad \delta = D_2/D_1, \quad \omega^2 = x_e/y_e \tag{8.192}$$

gives

$$\frac{\partial u}{\partial \tau} = \nabla^2 u + (1+u)v,$$
$$\frac{\partial v}{\partial \tau} = \delta\nabla^2 v - \omega^2(u+v+uv). \tag{8.193}$$

The linearization of these equations can be written

$$\frac{\partial \mathbf{w}}{\partial \tau} = \boldsymbol{\Delta}\nabla^2\mathbf{w} + \mathbf{K}\mathbf{w}, \tag{8.194}$$

where

$$\mathbf{w} = \begin{bmatrix} u \\ v \end{bmatrix}, \qquad \boldsymbol{\Delta} = \begin{bmatrix} 1 & \cdot \\ \cdot & \delta \end{bmatrix}, \qquad \mathbf{K} = \begin{bmatrix} \cdot & 1 \\ -\omega^2 & -\omega^2 \end{bmatrix}. \tag{8.195}$$

Since the matrix $\mathbf{K} - \lambda^2\boldsymbol{\Delta}$ is always negative definite, the analysis of Scriven and Othmer considered in Section 8.12 shows that the uniform steady state is stable and that travelling waves will be damped.

Frank-Kamenetskii (1969, p. 519) has suggested a similar scheme to explain cold flames, but with the reaction A → Y also autocatalytic. This replaces the term k_0 in the second eqn (8.201) by $k_0 y$, making $x_e = k_0/k_1$, and $y_e = k_2/k_1$. However the reduction to dimensionless form can be done with exactly the same variables and gives

$$\frac{\partial u}{\partial \tau} = \nabla^2 u + (1+u)v,$$
$$\frac{\partial v}{\partial \tau} = \delta\nabla^2 v - \omega^2(1+v)u. \tag{8.196}$$

The linearized equations for the uniform state now allow for a steady oscillation of frequency ω, but $\mathbf{K} - \lambda^2\boldsymbol{\Delta}$ is negative definite for all $\lambda > 0$, so that any non-uniform state will be damped.

Travelling bands of periodically reacting mixtures have been observed both in the well stirred bulk phase and in the shallow dishes and Zhabotinskii (1970) has given a qualitative explanation of some of the interactions of wave trains. Waves of spiral form have also been observed (cf. Winfree (1972)). Scriven, DeSimone, and Beil (1973) were however able to prepare membranes in which the progress of the Belousov–Zhabotinskii reaction could be studied without the unpredictable influence of convective effects. They found a great variety of wave-like phenomena and showed that spatial waves are possible if one of the steps is autocatalytic. Since one of the participating species is virtually immobilized in such a membrane, the simplest model is given by the equations

$$\frac{\partial x}{\partial \tau} = \nabla^2 x + f(x, y),$$
$$\frac{\partial y}{\partial \tau} = g(x, y). \tag{8.197}$$

If $f(x_e, y_e) = g(x_e, y_e) = 0$ the usual linearization about the steady state will give

$$\frac{\partial u}{\partial \tau} = \nabla^2 u + \alpha u + \beta v,$$
$$\frac{\partial v}{\partial \tau} = \gamma u + \delta v. \tag{8.198}$$

Unattenuated travelling plane waves may be found if equations admit a solution of the form

$$u(\rho, \tau) = U\,\mathrm{e}^{i\omega(\tau - \rho/\kappa)},$$
$$u(\rho, \tau) = V\,\mathrm{e}^{i\omega(\tau - \rho/\kappa)}, \tag{8.199}$$

where U and V are constants. This is possible if $\alpha + \delta > 0$ and $\beta\gamma < -\delta^2$ and then

$$\omega^2 = -(\beta\gamma + \delta^2), \qquad \kappa^2 = -(\beta\gamma + \delta^2)/(\alpha + \delta). \tag{8.200}$$

But this implies some degree of autocatalysis since α and δ cannot both be negative.

A spiral wave form will be given if

$$u(\rho, \tau) = U(\rho)\exp[i\{m\theta + \omega(\tau - \sigma\theta)\}],$$
$$v(\rho, \tau) = V(\rho)\exp[i\{m\theta + \omega(\tau - \sigma\theta)\}] \tag{8.201}$$

for such a wave form is periodic in θ with $(m - \omega\sigma)$ arms, and rotates with angular velocity $(1/\sigma)$. But if we substitute this in the eqns (8.198) we have

$U(\rho)$ proportional to the Bessel function $J_n(\rho\sqrt{(\alpha+\delta)})$ provided

$$\alpha+\delta > 0, \quad \beta\gamma < -\delta^2, \quad \omega^2 = -(\beta\gamma+\delta^2), \quad \sigma = (m-n)/\omega, \quad (8.202)$$

where m and n are integers. In contrast to the plane waves which are unattenuated, these spiral waves are asymptotically proportional to

$$\sqrt{\left(\frac{2}{\pi\rho}\right)}(\alpha+\delta)^{-\frac{1}{4}}\exp[i\{(\alpha+\delta)^{\frac{1}{2}}\rho+n\theta+\omega\tau\}]$$

and, as Scriven, DeSimone, and Beil point out, are geometrically attenuated by a factor of $\rho^{-\frac{1}{2}}$.

The Brussels school have studied systems with so-called dissipative structure and it would be beyond our scope to attempt to survey the whole of their work. The book of Glansdorff and Prigogine (1971) gives some account of their labours and a few references are given at the end of the chapter. The single example that must suffice here is, however, sufficiently striking. It is based on the kinetic scheme

$$\mathrm{A} \rightarrow \mathrm{X},$$
$$\mathrm{B}+\mathrm{X} \rightarrow \mathrm{Y}+\mathrm{D},$$
$$2\mathrm{X}+\mathrm{Y} \rightarrow 3\mathrm{X},$$
$$\mathrm{X} \rightarrow \mathrm{C},$$

in which A and B are supplied so as to keep their concentrations constant. In fact it can be regarded as a mechanism for the overall reaction $\mathrm{A}+\mathrm{B} \rightarrow \mathrm{C}+\mathrm{D}$ with the internal autocatalytic step $2\mathrm{X}+\mathrm{Y} \rightarrow 3\mathrm{X}$: it is not suggested that it corresponds to any known reaction, but may be regarded as a model reaction for a much more complicated system. Assuming that the kinetics are given by the stoicheiometry and that all of the rate constants are suitably adjusted the rates of change of x and y, the concentrations of X and Y, in a well-stirred reactor would be

$$\begin{aligned}\dot{x} &= a-(1+b)x+x^2y,\\ \dot{y} &= bx-x^2y.\end{aligned} \quad (8.203)$$

The steady state is

$$x_e = a, \qquad y_e = b/a \quad (8.204)$$

and the local linearization gives

$$\begin{aligned}\dot{u} &= -(1-b)u+a^2v,\\ \dot{v} &= bu-a^2v.\end{aligned} \quad (8.205)$$

It follows that the critical point is unstable if

$$b > 1+a^2 \quad (8.206)$$

(a)

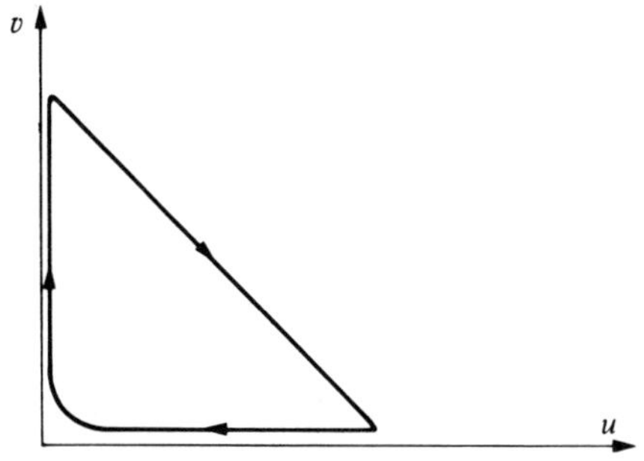

(b)

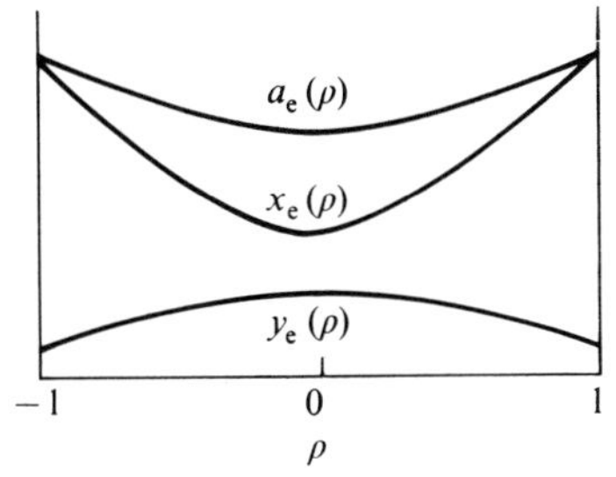

(c) $x(\rho)$

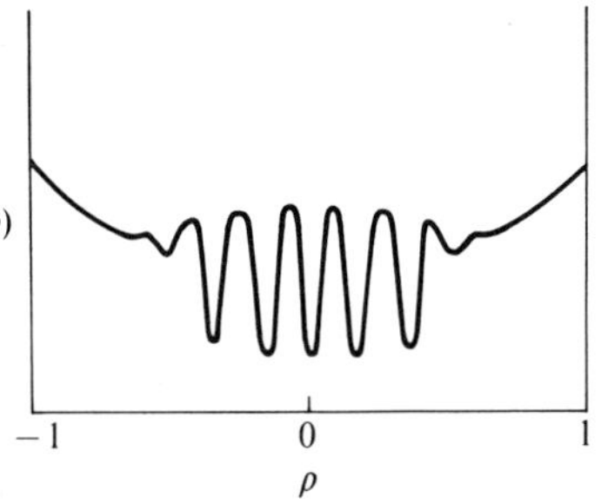

FIG. 8.14. (a) Limit cycle for the Herschkowitz-Kaufman reaction in stirred tank, (b) steady state profiles for low values of b, (c) confined periodic steady state for larger values of b. (After Herschkowitz-Kaufman and Nicolis (1972).)

and the system executes stable limit cycles of a highly nonlinear form. Herschkowitz-Kaufman, Nicolis and Lavenda (1971) have analysed these relaxation oscillations in detail and for certain values of the parameters the phase plane trajectories assume the almost triangular form shown in Fig. 8.14(a).

Herschkowitz-Kaufman, with Platten (1971) and with Nicolis (1972), has studied this system of kinetics when A is not held constant but is supplied to the region by diffusion and the species X and Y are also allowed to diffuse.

The equations governing $a(\rho, \tau)$, $x(\rho, \tau)$, and $y(\rho, \tau)$ would then be

$$\begin{aligned}\frac{\partial a}{\partial \tau} &= \alpha\frac{\partial^2 a}{\partial \rho^2} - a,\\ \frac{\partial x}{\partial \tau} &= \beta\frac{\partial^2 x}{\partial \rho^2} + a - (1+b)x + x^2 y,\\ \frac{\partial y}{\partial \tau} &= \gamma\frac{\partial^2 y}{\partial \rho^2} + bx - x^2 y,\end{aligned} \tag{8.207}$$

where the region of interest might be bounded by the planes $\rho = \pm 1$ on which the Dirichlet conditions

$$a = a_s, \qquad x = a_s, \qquad y = b/a_s \tag{8.208}$$

are imposed. In this case if $\alpha \to \infty$ so that the species A diffuses very rapidly, there is a steady state

$$a(\rho, \tau) = a_e(\rho) \equiv a_s, \qquad x(\rho, \tau) \equiv a_s, \qquad y(\rho, \tau) \equiv b/a_s.$$

When α is finite but large steady state of a has the usual form

$$a_e(\rho) = a_s\frac{\cosh(\rho\alpha^{-\frac{1}{2}})}{\cosh(\alpha^{-\frac{1}{2}})} \tag{8.209}$$

and the steady state forms of x and y are as sketched in Fig. 8.14(b). It might be expected that for a sufficiently large value of b this steady state would be unstable and indeed it is so. But the extraordinarily beautiful result revealed by the calculations of Herschkowitz-Kaufman and her colleagues is that a new steady state with a localized periodic structure is possible. This is shown qualitatively in Fig. 8.14(c) where it can be seen that the spatially periodic structure is confined to the central half of the region and the steady state outside this is close to the usual catenary form. This organization of the chemical species is maintained by the presence of B and the diffusion of A from the boundary, and its confinement can be explained by looking for the region in which a_e falls below a critical relationship with respect to the constant concentration b. Equation (8.206) is such a relation for the well-mixed reactor and Nicolis in his review paper (1971) has developed a stability criterion for the uniform steady state of eqns (8.207) with $\alpha = 0$. Thus if a and b are constant in the second and third of these equations and we set

$$x = a(1+u), \qquad y = (b/a)(1+v),$$

we can linearize the equations as

$$\begin{aligned}\frac{\partial u}{\partial \tau} &= \beta\frac{\partial^2 u}{\partial \rho^2}-(1-b)u+bv,\\ \frac{\partial v}{\partial \tau} &= \gamma\frac{\partial^2 v}{\partial \rho^2}-a^2u-a^2v.\end{aligned} \tag{8.210}$$

A solution of the form

$$u = U\exp(\omega\tau+i\rho/\lambda), \qquad v = V\exp(\omega\tau+i\rho/\lambda)$$

will only be possible if ω satisfies secular equation

$$\omega^2+\omega\left(\frac{\beta+\gamma}{\lambda^2}+1+a^2-b\right)+\left\{\left(1+\frac{\beta}{\lambda^2}\right)\left(\frac{\gamma}{\lambda^2}+a^2\right)-\frac{\gamma}{\lambda^2}b\right\} = 0. \tag{8.211}$$

This will have a non-oscillatory incipient stability if the roots are real and the last term vanishes, i.e.

$$b = \left(1+\frac{\beta}{\lambda^2}\right)\left(1+\frac{a^2\lambda^2}{\gamma^2}\right). \tag{8.212}$$

But this is satisfied for the smallest value of b if $\lambda^4 = \beta\gamma/a^2$ and the corresponding value of b would be

$$b = \{1+a(\beta/\gamma)^{\frac{1}{2}}\}^2. \tag{8.213}$$

We would therefore expect that a steady periodic structure might arise where

$$a_e(\rho) < \left(\frac{\gamma}{\beta}\right)^{\frac{1}{2}}(b^{\frac{1}{2}}-1).$$

This is the region

$$\rho < \alpha^{\frac{1}{2}}\cosh^{-1}\left\{\left(\frac{\gamma}{\beta}\right)^{\frac{1}{2}}\frac{b^{\frac{1}{2}}-1}{a_s}\cosh\alpha^{-\frac{1}{2}}\right\}. \tag{8.214}$$

Such a criterion is only approximate but gives an indication of how the confined periodic structure will arise.

For larger values of b the periodic steady state was found to be unstable and the concentrations of X and Y at each point executed limit cycles not unlike those found for eqns (8.203). This of course is reflected in the spatial distribution of the concentrations and waves very like the limit cycles we have observed in catalyst particles (Section 8.2) are developed. Figure 8.15 is a qualitative picture based on Herschkowitz-Kaufman and Nicolis' figures (1972) and is represented as a series of frames at a sequence of time intervals. From the initial catenary of the X concentration (Fig. 8.15(a)) a depression develops as the centre concentration falls (Fig. 8.15(b)). Concentration waves

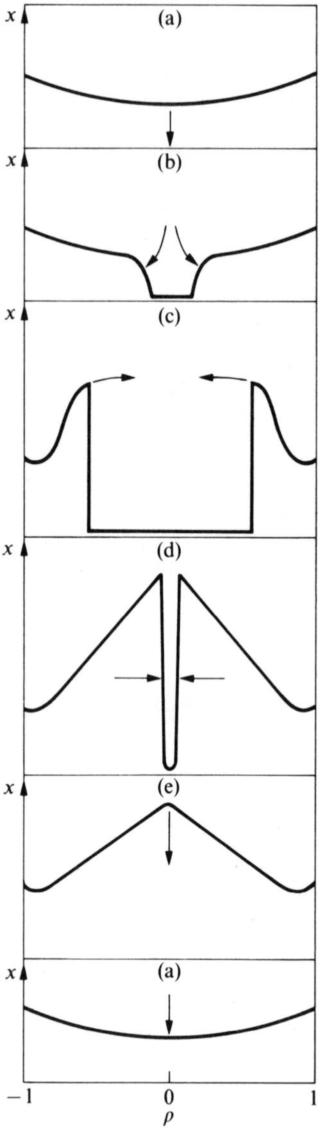

FIG. 8.15. Sequence of the spatial waves in the concentration of X for very large concentrations of B. (After Herschkowitz-Kaufman and Nicolis (1972).)

then spread out comparatively slowly, building up sharp peaks near the quarter and three-quarter points (Fig. 8.15(c)). There follows a rapid propagation of the two inner fronts toward the centre and then a slow subsidence of the central hump of concentration (Fig. 8.15(e)) to its original form, and the start of a new cycle. It is noteworthy that outer eighths of the profiles are virtually unaffected by the violent oscillations within.

This example, alone would serve to show what an extraordinarily rich region of the domain of diffusion and reaction awaits to be more fully explored. Let it stand as a tribute to the pioneers who have worked their way into the whole domain from so many directions, a beacon to those who would follow and go beyond them and a colophon to this all-too-inadequate account of some of the explorations that have already been made.

Additional bibliographical comments

8.1. Kasaoka, Sakata, and Nitta (1968) show that the shape of particles becomes of minor importance when the characteristic dimension used in making the variables dimensionless is V_p/S_x i.e. $\upsilon = \sigma$. Prager and Frisch (1971) have considered the time lag in establishing the steady flux through a wall and have shown that it can be expressed in terms of the final steady state even in an inhomogeneous medium. Some of the first diffusion and reaction problems that were solved included the transient case cf. Jüttner (1909) and Roughton (1952). For more on the relation between problems with and without reaction see Danckwerts (1951), Stewart (1968), Pao (1964, 1965), Rosen (1973*a*).

8.2. For the study of the regimes of the solution to the stirred tank equations see Hlaváček, Kubíček, and Jelínek (1970), Hlaváček, Kubíček, and Višňák (1972). A good general review is to be found in Ray (1972*a*), but the definitive survey for the stirred tank is to be found in Poore, Uppal, and Ray (1974). Kochina (1968) has discussed limit cycles that arise from the boundary conditions. For transient measurements see Benham and Denny (1972). Bidner and Calvelo (1974) have performed calculations for the transient Robin problem with $\mathscr{L} = 1$ and have shown how the regions of attraction of the several steady states are disposed.

8.4. Bentwich (1971, 1973) has also used singular perturbation methods for diffusion reaction problems in which the reaction rate is small compared with the transport rate. See also Cole (1968) for a description of the method of two time scales. The bifurcation of limit cycles at a transition of stability of the stirred tank was first discussed by Amundson and Aris (1956). See also Lapidus and Luus (1966, 1967), Friedly (1972), Douglas (1972), Keller (1970), Keller and Keener (1973).

8.5. Other examples of the value of periodic operation are given in the work of Douglas and his colleagues: for references see Douglas (1972), Friedly (1972).

8.6. See also Lions (1971) for a comprehensive, if somewhat abstract, discussion of optimal control problems.

8.7. A number of references to the subject of catalyst deactivation are given in the bibliography for Section 5.1.

8.9. For a discussion of the finite propagation speed of heat and matter see Chester (1963), Weymann (1967), Gurtin and Pipkin (1968), Chen (1969), and Bauermeister and Hamill (1969). The last of these refers to the problem of the inception of nucleate boiling which may justify the use of the telegrapher's equation. For the derivation of this equation from kinetic theory see Dahler and Sandler (1964). Ruckenstein (1972) has discussed also the stability of supported clusters, from a thermodynamic point of view, and the mechanism of decay (Ruckenstein and Pulvermacher 1972).

8.10. For further discussion of the Kolmogoroff–Petrovsky–Piscounoff problem see Il'in and Oleinik (1958), Kanel (1960, 1961), Zelenyak (1966, 1967) and references given in Zel'dovich and Barenblatt (1971). See also Fisher (1937), Itô and McKean (1965), Montroll (1967), Cohen (1970), and Rosen (1973*d*). On the Belousov–Zhabotinskii reaction see Zhabotinskii (1964*a*, 1967, 1970) and his papers with Zaikin, Korzukhin and Kreitser (1971) and Deshcherevskii, Sel'kov, Sidorenko, Shnol' (1970). See also Degn (1967), Franck (1967), Kasparek and Bruice (1971), Field and Noyes (1972), Winfree (1972), Scriven, DeSimone, and Beil (1973), Kopell and Howard (1973*a*, *b*), and Rosen (1973*b*, *c*). Frank-Kamenetskii touches on these topics in his second edition of 1969. There is much on dissipative structures in Glansdorff and Prigogine (1971). See also Prigogine and Nicolis (1967), Edelstein (1970), Herschkowitz-Kaufman (1970) and her papers with Nicolis (1972), Nicolis and Lavenda (1971) and Platten (1971). A very useful review is that of Nicolis (1971). Segel and Jackson give a lucid explanation of how instability arises in their 1972 paper. For general considerations on oscillating reactions, see Higgins (1967). The conditions under which a network of chemical reactions can give rise to limit cycles and other oscillatory phenomena are a by-product of the analysis of the algebraic structure of kinetics to be found in the work of Feinberg (1972, 1973), Horn (1973), and Jackson (Horn and Jackson, 1972). Ortoleva and Ross (1972*a*) have used the model reaction of the Brussels school to illustrate their analysis of the penetration of boundary perturbations into unstable chemical systems. They have also discussed the use that might be made of such perturbations (1972*b*) and developed a theory of phase waves which accords with some of the observations on the Belousov–Zhabotinskii reaction (1973). See also Kopell and Howard (1973*a*, *b*), Tatterson and Hudson (1973).

Martinez and Baer (1973) have shown how development can proceed in a linear array of cells with such a reaction scheme as that of Herschkowitz-Kaufman. If a steady state, which is stable in isolated cells, is no longer stable when two cells are brought into contact and can exchange reactants, then an asymmetrical steady state may be the only stable one. These authors then assume that mitosis takes place when $x-y > c$ in a given cell and that this produces two equal daughter cells of the same constitution as the parent. If this arrangement is still unstable it leads to a different pattern of steady states within which further mitosis is possible and hence to a developing, patterned array of cells.

NOMENCLATURE

(See also the General Nomenclature)

A, B, C, D	reactants
A	matrix of coefficients in discretized equations, Section 8.3
$A(t), B(t), C(t)$	amplitude coefficients, $A^2+B^2 = C^2$, Section 8.4
a, b, c, d	concentrations of A, B, C, D

a_n	coefficients in eigenfunction expansion, Section 8.1
C_p	heat capacity of particle
$\hat{c}$	adsorbed concentration
c	speed of propagation of temperature
E	parameter Section 8.8
$\mathscr{G}$	Green's function for eqn (8.50)
$\mathbf{G}$	exponential matrix in Section 8.3
H	Heaviside step function, Section 8.9
$J(V)$	cost functional, Section 8.6
$\mathbf{K}$	matrix of kinetic constants
k_a, k_d	adsorption and desorption rate constants
$\mathscr{L}_c$	critical Lewis number
L	analogue of Lewis number for the stirred tank
$L_c, \hat{L}_c$	critical Lewis number for stirred tank without and with adsorption
P_i	production rates, Section 8.5
p, q	adjoint variables, Section 8.6
q	heat released by reaction, Section 8.9
R_{pi}	$(i = 1, 2, 3)$, rate of poisoning reaction of type i
s	Laplace transform variable
T, t	time scales in Section 8.4
t_p	period of heat release by reaction, Section 8.9
t_r	duration of heat release by reaction, Section 8.9
$u_e(\boldsymbol{\rho}, \phi^2, \nu)$	steady state solution of eqns (8.1), (8.2) with parameters ϕ, ν
U_n, V_n, u_n, v_n	terms in perturbation expansion
$V(\tau), V^0(\tau)$	controlled concentration of v_s, optimal value Section 8.6
$W(\tau)$	total solute taken up in dyeing problem, Section 8.1
w_n	eigenfunctions of eqn (8.6)
X, Y	intermediates
x, y	concentrations of X, Y : transformed variables in eqn (8.78)
α	conductivity
α^2	$R'(0)$ in Section 8.10
α, β, γ	reciprocal Thiele moduli in Section 8.10
$\alpha(\lambda), \beta(\lambda)$	coefficients in eqn (8.79)
$\boldsymbol{\Delta}$	diffusivity matrix
δ	diffusivity ratio
$\hat{\varepsilon}$	adsorption capacity parameter Section 8.8
ζ	$\xi + \kappa\tau + \mu$
κ	wave velocity in Section 8.10
κ, λ, μ	kinetic parameters in Section 8.5
λ_n	eigenvalue for eqn (8.6)
Λ	dimensionless thickness, Section 8.9

μ', ν'	$\mu/\phi^2, \nu/\phi^2$
τ'	dimensionless time defined by eqn (8.59)
τ_{h}	relaxation time for heat, α/c^2
ω	frequency
ψ	analogue of the Thiele modulus for stirred tank
$\bar{\cdot}$	denotes the Laplace transform
$\langle \cdot \rangle$	average over Ω
$\tilde{\cdot}$	analogous variable for the stirred tank
$\cdot^*$	deviation from steady state.

REFERENCES

I HAVE occasionally taken the liberty of changing the order of the authors' names from that in which they stood in the original paper so as to bring together work that was done in the same school or under the same leadership. I trust the displaced authors will not mind if the reader has at times to use the index to locate them. Titles of the papers have been given in full as laying the sinews, if not the flesh and skin, on what otherwise are but dry bones. I have referred to certain Ph.D dissertations since these are available in microfilm from University Microfilms, Ann Arbor, Michigan. Unless otherwise stated, the paper is in the language of the title.

AMANN, H. (1971). On the existence of positive solutions of nonlinear elliptic boundary value problems. *Indiana Univ. Publs Sci. Ser. Maths Section* **21**, 125.

———. (1972*a*) A uniqueness theorem for nonlinear elliptic boundary value problems. *Arch. ration Mech Analysis* **44**, 178.

———. (1972*b*). Existence of multiple solutions for nonlinear elliptic boundary value problems. *Indiana Univ. Publs Sci. Ser., Maths Section* **21**, 925.

———. (1972). On the number of solutions for nonlinear equations in ordered Banach spaces. *J. Fnl. Anal.* **11**, 346.

AMUNDSON, N. R. (1965). Some further observations on tubular reactor stability. *Can. J. chem. Engng* **43**, 49.

——— and ARIS, R. (1958). An analysis of chemical reactor stability and control-II. The evolution of proportional control. *Chem. Engng Sci.* **7**, 132.

——— and KUO, J. C. W. (1967*a*). Catalytic particle stability studies. I. Lumped resistance model. *Chem. Engng Sci.* **22**, 49.

———, ———. (1967*b*). Catalytic . . . II. Lumped thermal resistance model. *Chem. Engng Sci.* **22**, 443.

———, ———. (1967*c*). Catalytic . . . III. Complex distributed resistances model. *Chem. Engng Sci.* **22**, 1185.

——— and LUSS, D. (1967*a*). Some general observations on tubular reactor stability. *Can. J. chem. Engng* **45**, 341.

———, ———. (1967*c*). Uniqueness of the steady state for an isothermal porous catalyst. *Ind. Eng. Chem., Fundam.* **6**, 457.

———, ———. (1967*d*). Uniqueness of the steady state solutions for chemical reaction occurring in a catalyst particle or in a tubular reactor with axial diffusion. *Chem. Engng Sci.* **22**, 253.

——— and MARKUS, L. (1968). Non-linear boundary-value problems arising in chemical reactor theory. *J. Different. Eqat.* **4**, 102.

——— and RAYMOND, L. R. (1964). Some observations on tubular reactor stability. *Can. J. chem. Engng* **42**, 173.

———, ———. (1965). Stability in distributed parameter systems. *A.I.Ch.E. Jl.* **11**, 339.

———, SCHNEIDER, D. R., and ARIS, R. (1973). An analysis of chemical reactor stability and control.—XIV. The effect of the steady state hypothesis. *Chem. Engng Sci.* **28**, 885.

AMUNDSON, N. R. and VARMA, A. (1972*a*). Global asymptotic stability in distributed parameter systems: comparison function approach. *Chem. Engng Sci.* **27**, 907.

———, ———. (1972*b*). Some problems concerning the non-adiabatic tubular reactor. *Can. J. chem. Engng* **50**, 470.

———, ———. (1973). Maximal and minimal solutions, effectiveness factors for chemical reactions in porous catalysts. *Chem. Engng Sci.* **28**, 91.

ARIS, R. (1969*b*). On stability criteria of chemical reaction engineering. *Chem. Engng Sci.* **24**, 149.

———. (1971*b*). A note on the structure of the transient behaviour of chemical reactors. *Chem. Eng. J.* **2**, 140.

———. (1972*b*). Some problems in the analysis of transient behaviour and stability of chemical reactors. Proc. 1st International Symposium on Chemical Reaction Engineering, Washington, 1970. Published as *Adv. Chem. Ser.* (Ed. R. F. Gould). **109**. American Chemical Society, Washington, 1972.

——— and COPELOWITZ, I. (1970*a*). Communications on the theory of diffusion and reaction—V. Findings and conjectures concerning the multiplicity of solutions. *Chem. Engng Sci.* **25**, 906.

——— and DROTT, D. W. (1969). Communications . . . I. A complete parametric study of the first-order, irreversible exothermic reaction in a flat slab of catalyst. *Chem. Engng Sci.* **24**, 541.

——— and GEORGAKIS, C. (1973). Communications . . . X. A generalization of Wei's bounds on the maximum temperature. *Chem. Engng Sci.* **28**, 291.

——— and REGENASS, W. (1965). Stability estimates for the stirred tank reactor. *Chem. Engng Sci.* **20**, 60.

ARONSON, D. G. and PELETIER, L. (1974). Global stability of symmetric and asymmetric profiles in catalyst particles. *Archs Ration Mech. Analysis* **54**, 175.

AUCHMUTY, J. F. G. (1973). Liapounov methods and equations of parabolic type. Proc. Battelle Res. Cen. 1972 Summer Institute *Nonlinear Mathematics*. Springer-Verlag, Heidelberg.

AULT, J. W. and MELLICHAMP, D. A. (1972). A diffusion and reaction model for simple polycondensation. *Chem. Engng Sci.* **27**, 1441.

BAHTIA, N. P. and SZEGO, G. P. (1970). *Stability theory of dynamical systems*. Springer-Verlag, Heidelberg.

BAILEY, J. E. (1969). *Improvement of catalytic selectivity by periodic operation*. Ph.D. Thesis, Rice University.

——— and LEE, C. K. (1974). Diffusion waves and selectivity modifications in cyclic operation of a porous catalyst. *Chem. Engng Sci.* **29**, 1157.

BAUERMEISTER, K. J. and HAMILL, T. D. (1969). Hyperbolic heat-conduction equation. A solution for the semi-infinite body problem. *J. Heat Transfer* **91**, 543.

BENHAM, C. B. and DENNY, V. E. (1972). Transient diffusion of heat, mass species and momentum in cylindrical pellets during catalytic oxidation of CO. *Chem. Engng Sci.* **27**, 2163.

BENTWICH, M. (1971). Singular perturbation solution of time dependent mass transfer with non-linear chemical reaction. *J. Inst. Math. its Appl.* **7**, 228.

———. (1973). Combined heat and mass transfer for two reacting species with temperature dependent rate constant. *Chem. Engng Sci.* **28**, 1465.

BESKOV, V. S. and MALINOVSKAYA, O. A. (1968). Stability and sensitivity of processes on porous pellets of catalyst. *IVth Int. Congr. Catal.*, Moscow.

BIDNER, M. S. and CALVELO, A. (1974). Transient analysis of exothermal reactions within catalyst pellets. Effect of initial conditions. *Chem. Engng Sci.* **29**, 1237.

BISCHOFF, K. B. (1963). Accuracy of the pseudo-steady-state approximation for moving boundary diffusion problems. *Chem. Engng Sci.* **18**, 711.

———. (1965*b*). Further comments on the pseudo-steady-state approximation for moving boundary diffusion problems. *Chem. Engng Sci.* **20**, 783.

BRUSSET, H., DEPEYRE, D., BOEDA, M., and HOANG ANH, V. U. (1972). Dimensionalment des grains de catalysateur pour assurer la stabilité d'un réacteur adiabatique. *Chem. Engng Sci.* **27**, 1475.

BUTT, J. B. (1972). Catalyst deactivation. Proc. 1st International Symposium, on Chemical Reaction Engineering, Washington 1970. Published as *Adv. Chem. Ser.* **109**, p. 259. American Chemical Society, Washington 1972.

——— and KEHOE, J. P. G. (1972*b*). Uniqueness of the steady state solutions in a catalyst pellet with a bi-modal pore size distribution. *Chem. Sci. Engng* **27**, 650.

CARSLAW, H. S. and JAEGER, J. C. (1959). *Conduction of heat in solids* (Second Edn). Clarendon Press, Oxford.

CHAN, S. H., LOW, M. J. D., and MUELLER, W. K. (1971). Hyperbolic heat conduction on catalytic supported crystallites. *A.I.Ch.E. Jl* **17**, 1499.

CHANDRASEKHAR, S. (1939). *Introduction to the study of stellar structure*. University of Chicago Press, Chicago.

CHEN, P. J. (1969). On the growth and decay of one-dimensional temperature rate waves. *Archs ration Mech. Analysis* **35**, 1.

CHESTER, M. (1963). Second sound in solids. *Phys. Rev.* **131**, 2013.

CHUA, L. O. and ALEXANDER, G. R. (1971). The effects of parasitic reactances on nonlinear networks. *IEEE Trans. Circuit Theory.* **CT-18**, 520.

COHEN, D. S. (1971). Multiple stable solutions of nonlinear boundary value problems arising in chemical reactor theory. *SIAM J. Appl. Math.* **20**, 1.

———. (1972). Multiple solutions of singular perturbation problems. *SIAM J. Numer. Anal.* **3**, 72.

———. (1973*a*). Multiple solutions and periodic oscillations in nonlinear diffusion processes. *SIAM J. Appl. Math.* **25**, 640.

———. (1973*b*). Multiple solutions of nonlinear partial differential equations. Proc. Battelle Res. Cen. 1972 Summer Institute. *Nonlinear Mathematics*. Springer-Verlag, Heidelberg.

——— and KELLER, H. B. (1967). Some positive problems suggested by nonlinear heat generation. *J. Math. Mech.* **16**, 1361.

——— and LAETSCH, T. W. (1970). Nonlinear boundary value problems suggested by chemical reactor theory. *J. Different Eqat.* **7**, 217.

——— and POORE, A. B. (1973). Multiple stable and unstable periodic solutions of the nonlinear equations of reactor theory.

COHEN, H. (1971). Nonlinear diffusion problems. *Studies in applied mathematics*, (Ed. A. H. Taub). American Mathematics Society and Prentice-Hall, New Jersey.

COLE, J. D. (1968). *Perturbation methods in applied mathematics*. Blaisdell, Waltham, Massachusetts.

CRANK, J. (1956). *The mathematics of diffusion*. Clarendon Press, Oxford.

CRESSWELL, D. L. (1970). On the uniqueness of the steady state of a catalyst pellet involving both intraphase and interphase transport. *Chem. Engng Sci.* **25**, 267.

——— and ELNASHAIE, S. S. E. (1973*a*). On the dynamic modelling of porous catalyst particles. *Can. J. chem. Engng* **51**, 201.

———, ———. (1973*b*). Dynamic behaviour and stability of non-porous catalyst particles. *Chem. Engng Sci.* **28**, 1387.

——— and PATERSON, W. R. (1971). A simple method for the calculation of effectiveness factors. *Chem. Engng Sci.* **26**, 605.

CRICK, F. H. C. (1970). Diffusion in embryogenesis. *Nature*. **225**, 420.

CRONIN, J. (1964). *Fixed points and topological degree in nonlinear analysis*. American Mathematics Society, Providence.

DAHLER, J. S. and SANDLER, S. I. (1964). Nonstationary diffusion. *Physics Fluids* **7**, 1743.

DANCKWERTS, P. V. (1951). Absorption by simultaneous diffusion and chemical reaction into particles of various shapes and into falling drops. *Trans. Faraday Soc.* **47**, 1014.

DEGN, H. (1967). Effect of bromine derivatives of malonic acid on the oscillating reaction of malonic acid, cerium ions and bromate. *Nature* **213**, 539.

———. (1968). Bistability caused by substrate inhibition of peroxidate in an open reaction system. *Nature* **217**, 1047.

DENN, M. M. (1970). A macroscopic condition for stability. *A.I.Ch.E. Jl* **16**, 670.

———. (1973). Stability of a catalytic reaction to finite amplitude disturbances. *Chem. Eng. J.* **4**, 105.

———. (1974). *Stability of reaction and transport processes*. Prentice-Hall, Englewood Cliffs, New Jersey.

DENTE, M. and BIARDI, G. (1967). Valutazione approssimata del campo di soluzioni multiple per una reazione esotermica evolventesi entro granuli. *Quad. Ing. Chim. Ital.* **3**, 185.

———, ———, and RANZI, E. (1969*a*). Un nuovo methodo per il calcolo dei profili di concentrazione e del fattore di utilizzazione per reazioni complesse evolventise i granuli catalitici non isothermi. *Quad. Ing. Chim. Ital.* **5**, 65.

———, ———, ———. (1969*b*). Problemi riguardanti la determinazione diretta dei limiti statici di 'ignizione' e 'spegnimento' per reazioni cineticamente semplici evolventisi in granuli catalitici non isotermi. *Quad. Ing. Chim. Ital.* **5**, 122.

DOUGLAS, J. M. (1972). *Process dynamics and control*. Prentice-Hall, Englewood Cliffs, New Jersey.

——— and YU, K. M. (1973). Self-generated oscillations on catalyst particles. *Chem. Engng Sci.* **28**, 163.

ECKHAUS, W. (1965). *Studies in nonlinear stability theory*. Springer-Verlag, Heidelberg.

EDELSTEIN, B. B. (1970). Instabilities associated with dissipative structure. *J. theor. Biol.* **26**, 227.

FEINBERG, M. (1972). On chemical kinetics of a certain class. *Archs ration Mech. Analysis* **46**, 1.

———. (1973). Complex balancing in general kinetic systems. *Archs ration Mech. Analysis* **49**, 187.

FIELD, R. J. and NOYES, R. M. (1972). Explanation of spatial band propagation in the Belousov reaction. *Nature* **237**, 390.

FINLAYSON, B. A. (1972). *The method of weighted residuals and variational principles.* Academic Press, New York.

——— and FERGUSON, N. B. (1971). Transient chemical reaction analysis by orthogonal collocation. *Chem. Eng. J.* **1**, 327.

FISHER, R. A. (1937). The wave of advance of advantageous genes. *Ann. Eugen.* **7**, 335.

FITZER, E. (1969). Dynamische Instabilitaten bei heterogenen Gas/Feststoff Reaktionen. *Chemie-Ingr.-Tech.* **41**, 331.

FJELD, M. (1967). *On the stability of distributed parameter systems.* Institutt for Reguleringsteknikk, N.T.H., Trondheim.

FRANCK, U. F. (1967). Phänomene on biologischen und knostlichen Membraner. *Ber. (Dtsch.) Bungenges phys. Chem.* **71**(8), 789.

FRANK-KAMENETSKII, D. A. (1969). *Diffusion and heat transfer in chemical kinetics.* (Second enlarged and revised edn.) Plenum Press, New York. (Nauka Press, Moscow 1967.)

FRIEDLY, J. C. (1972). *Dynamic behavior of processes.* Prentice-Hall, Englewood Cliffs, New Jersey.

FRIEDMAN, A. (1966). *Partial differential equations of parabolic type.* Prentice-Hall, Englewood Cliffs, New Jersey.

FUJITA, H. (1969). On the nonlinear equations $\Delta u + e^u = 0$ and $\partial v/\partial t = \Delta v + e^v$. *Am. math. Mon.* **76**, 132.

GALL, C. E. (1972). Nonlinear functional analysis and reactor stability. Proc. 1st International Congress on Chemical Reaction Engineering, Washington 1970. Published as *Adv. Chem. Ser.* (Ed. R. F. Gould) **109**, p. 622. American Chemical Society, Washington.

GAVALAS, G. R. (1966). On the steady states of distributed parameter systems with chemical reactions, heat and mass transfer. *Chem. Engng Sci.* **21**, 477.

———. (1967). The behaviour of distributed systems with complex reactions and diffusion in the regime of transport limitation. *Chem. Engng Sci.* **22**, 997.

———. (1968). *Nonlinear differential equations of chemically reacting systems.* Springer-Verlag, Heidelberg.

GELFAND, I. M. (1959). Some problems in the theory of quasilinear equations. *Uspehi. mat. Nauk* **14**, No. 2(86), 87. English Translation: *Am. math. Soc. Transl.* (2) **29** (1963), 295.

GEORGAKIS, C. and SANI, R. L. (1974). On the stability of the steady state in systems with coupled diffusion and reaction. *Arch. ration Mech. Analysis* **52**, 266.

GILLES, E. D. (1964). Methode zur Berechnung stationärer Grenzschwingungen in Rührkesselreaktoren. *Grundlagen der chemischen Prozessregelung* (Ed. W. Oppelt and E. Wicke). R. Oldenbourg, Verlage, Munchen.

GLANSDORFF, P. and PRIGOGINE, I. (1971). *Thermodynamic theory of structure, stability and fluctuations.* Wiley, New York.

GMITRO, J. I. (1969). *Concentration patterns generated by reaction and diffusion.* Ph.D. Thesis, University of Minnesota.

GRAY, P. and HARPER, M. J. (1959). The thermal theory of induction periods and ignition delays. *VIIth Int. Symp. Combust.* p. 425. Butterworths, London.

GURTIN, M. E. and PIPKIN, A. C. (1968). A general theory of heat conduction with finite wave speeds. *Archs ration Mech. Analysis* **31**, 113.

HANSEN, K. W. (1971). Analysis of transient models for catalytic tubular reactors by orthogonal collocation. *Chem. Engng Sci.* **26**, 1555.

HELLINCKX, L., GROOTJANS, J., and VAN DEN BOSCH, B. (1972). Stability analysis of the catalyst particle through orthogonal collocation. *Chem. Engng Sci.* **27**, 644.

HERSCHKOWITZ-KAUFMAN, M. (1970). Structures dissipatives dans une réaction chimique homogène. *C.R. hebd. Séanc. Acad. Sci., Paris.* **270**, 1049.

——— and NICOLIS, G. (1972). Localized spatial structures and nonlinear chemical waves in dissipative systems. *J. chem. Phys.* **56**, 1890.

———, ——— and LAVENDA, B. (1971). Chemical instabilities and relaxation oscillations. *J. theor. Biol.* **32**, 283.

——— and PLATTEN, J. K. (1971). Chemical instabilities and localized structures in non-homogeneous media. *Bull. Acad. r. Belg. Cl. Sci.* **57**, 26.

HIGGINS, J. (1967). The theory of oscillating reactions. *Applied kinetics and chemical reaction engineering*. (Ed. R. L. Gorring and V. W. Weekman). American Chemical Society, Washington.

HLAVÁČEK, V. (1970*a*). Modeling and stability of catalytic fixed bed reactors. *Can. J. chem. Engng* **48**, 656.

———. (1970*b*). Aspects in design of packed catalytic reactors. *Ind. Eng. Chem.* **62** (7), 8.

——— and HOFMANN, H. (1970*a*). Modelling of chemical reactors. XVI. Steady state axial heat and mass transfer in tubular reactors. *Chem. Engng Sci.* **25**, 173.

———, ———. (1970*b*). Modelling . . . XVII. Steady state axial heat and mass transfer in tubular reactors. *Chem. Engng Sci.* **25**, 187.

———, ———. (1970*c*). Modelling . . . XIX. Transient axial heat and mass transfer in tubular reactors. The stability considerations—I. *Chem. Engng Sci.* **25**, 1517.

——— and KUBÍČEK, M. (1970*a*). Modelling . . . XX. Heat and mass transfer in porous catalyst. The particle in a non-uniform external field. *Chem. Engng Sci.* **25**, 1527.

———, ———. (1970*b*). Modelling . . . XXI. Effect of simultaneous heat and mass transfer inside and outside a pellet on reaction rate—I. *Chem. Engng Sci.* **25**, 1537.

———, ———. (1970*c*). Modelling . . . XXII. Effect of simultaneous heat and mass transfer inside and outside of a pellet on reaction rate—II. *Chem. Engng Sci.* **25**, 1761.

———, ———. (1971*a*). Modelling . . . XXIII. Transient heat and mass transfer in a porous catalyst. III. Reaction with finite values of Sherwood and Nusselt number. Effect of volume contraction. *J. Catal.* **22**, 364.

———, ———. (1971*b*). Qualitative analysis of the behaviour of nonlinear parabolic equations.—I. Development of methods. *Chem. Engng Sci.* **26**, 1737.

———, ——— and CAHA, J. (1971). Qualitative analysis . . . II. Application of the methods for the estimation of domains of multiplicity. *Chem. Engng Sci.* **26**, 1743.

———, ———, and JELÍNEK, J. (1970). Modelling . . . XVIII. Stability and oscillatory behaviour of the CSTR. *Chem. Engng Sci.* **25**, 1444.

———, ———, and VIŠŇÁK, K. (1972). Modelling . . . XXVI. Multiplicity and stability analysis of a continuous stirred tank reactor with exothermic consecutive reactions A → B → D. *Chem. Engng Sci.* **27**, 719.

——— and MAREK, M. (1968*a*). Modelling . . . VI. Heat and mass transfer in a porous catalyst particle. On the multiplicity of solutions for the case of an exothermic zeroth-order reaction. *Colln. Czech. chem. Commun. Engl. Edn* **33**, 506.

———, ———, and KUBÍČEK, M. (1968*b*). Modelling . . . X. Multiple solutions of enthalpy and mass balances for a catalytic reaction within a porous catalyst particle. *Chem. Engng Sci.* **23**, 1083.

———, ———, ———. (1969*a*). Modelling . . . XIV. Analysis of nonstationary heat and mass transfer in a porous catalyst particle.—I. *J. Catal.* **15**, 17.

———, ———, ———. (1969*b*). Modelling . . . XV. Analysis of nonstationary heat and mass transfer in a porous catalyst particle. II. *J. Catal.* **15**, 31.

———, ———, ———. (1970). Modelling . . . XIII. Non stationary heat and mass transfer in a particle of porous catalyst. One dimensional approximation of the model. V. *Colln. Czech. chem. Commun. Engl. Edn* **35**, 2124.

HORN, F. J. M. (1973). Necessary and sufficient conditions for complex balancing in chemical kinetics. *Archs ration Mech. Analysis* **49**, 172.

——— and BAILEY, J. E. (1968). An application of the theorem of relaxed control to the problem of increasing catalyst selectivity. *J. Optiz. Theory Appl.* **2**, 441.

———, ———. (1969). Catalyst selectivity under steady-state and dynamic operation. *Ber. (Dtsch.) Bunsenges phys. Chem.* **73**, 274.

———, ———. (1970). Catalyst selectivity under steady-state and dynamic operation: an investigation of several kinetic mechanisms. *Ber. (Dtsch.) Bunsenges phys. Chem.* **74**, 611.

———, ———. (1972). Cyclic operation of reaction systems: the influence of diffusion on catalyst selectivity. *Chem. Engng Sci.* **27**, 109.

——— and JACKSON, R. (1972). General mass action kinetics. *Archs ration Mech. Analysis* **47**, 81.

IL'IN, A. M., KALASHNIKOV, A. S., and OLEINIK, O. A. (1962). Second-order linear equations of parabolic type. *Russ. math. Survs* **17(3)**, 1.

——— and OLENIK, O. A. (1958). On the behaviour of solutions of the Cauchy problem for certain quasi-linear equations under unbounded increase of time. *Dokl. Akad. Nauk SSSR* **120**, 1. (In Russian.)

ITO, K. and MCKEAN, H. P. (1965). *Diffusion processes and their sample paths*. Springer-Verlag, Heidelberg.

JACKSON, R. (1972*a*). The stability of standing waves on a catalytic wire. *Chem. Engng Sci.* **27**, 2304.

———. (1972*b*). Some uniqueness conditions for the symmetric steady state of a catalyst particle with surface resistances. *Chem. Engng Sci.* **27**, 2205.

———. (1973). A simple geometric condition for instability in catalyst pellets at unit Lewis number. *Chem. Engng Sci.* **28**, 1355.

JOSEPH, D. D. (1964). Variable viscosity effects on the flow and stability of flow in channels and pipes. *Physics Fluids* **7**, 1761.

JOSEPH, D. D. (1965). Non-linear heat generation and stability of the temperature distribution in conducting solids. *Int. J. Heat Mass Transfer* **8**, 281.

———. (1966). Bounds on λ for positive solutions of $\Delta\psi+\lambda f(r)\{\psi+G(\psi)\}=0$. *Q. appl. Math.* **23**, 349.

——— and LUNDGREN, T. S. (1973). Quasilinear Dirichlet problems driven by positive sources. *Arch. Ration Mech. Analysis* **49**, 241.

——— and SPARROW, E. M. (1970). Nonlinear diffusion induced by nonlinear sources. *Q. appl. Math.* **28**, 327.

KAGANOV, S. A. (1963). Establishing laminar flow for an incompressible liquid in a horizontal channel or curved cylindrical tube with corrections for frictional heat and the temperature dependence of viscosity. *Int. chem. Eng.* **3**, 33.

KANEL, Y. I. (1960). The behavior of solutions of the Cauchy problem when the time tends to infinity in the case of quasilinear equations arising in the theory of combustion. *Dokl. Akad. Nauk SSSR* **132**, 268. (Translated from *Sov. Math.* **1**, 533.)

———. (1961). Some problems involving burning-theory equations. *Sov. Math.* **2**, 48.

KASAOKA, S., SAKATA, Y., and NITTA, K. (1968). Unsteady state diffusion within porous solid particles. Correlation between solid shapes and diffusion rates. *J. chem. Engng, Jap.* **1**, 32.

KASPAREK, G. J. and BRUICE, T. C. (1971). Observations on an oscillation reaction. The reaction of potassium bromate, ceric sulfate and a dicarboxylic acid. *Inorg. Chem.* **10**, 382.

KASTENBERG, W. E. (1967). On the stability of equilibrium of a distributed parameter feedback control system. *Int. J. Contr.* **6**, 523.

———. (1969). A stability criterion for space-dependent nuclear-reactor systems with variable temperature feedback. *Nucl. Sci. Engng* **37**, 19.

———. (1970). Comparison theorems for nonlinear multicomponent diffusion systems. *J. math. Analysis Applic.* **29**, 299.

———. (1973). On global stability of distributed parameter chemical reaction systems. *Chem. Engng Sci.* **28**, 1691.

——— and CHAMBRE, P. L. (1968). On the stability of nonlinear space-dependent reactor kinetics. *Nucl. Sci. Engng* **31**, 67.

——— and ZISKIND, R. A. (1974). On the stability of diffusion systems with chemical reactions. *Math. Biosci.* To appear.

KELLER, H. B. (1969*a*). Positive solutions of some nonlinear eigenvalue problems. *J. Math. Mech.* **19**, 279.

———. (1969*b*). Elliptic boundary value problems suggested by nonlinear diffusion processes. *Archs ration Mech. Analysis* **35**, 363.

———. (1970). Nonlinear bifurcation. *J. Different Eqat.* **7**, 417.

——— and KEENER, J. P. (1973). Perturbed bifurcation theory. *Archs ration Mech. Analysis* **50**, 159.

——— and LANGFORD, W. F. (1972). Iterations, perturbations and multiplicities for non-linear bifurcation problems. *Archs ration Mech. Analysis* **48**, 83.

KESTEN, A. S. and SANGIOVANNI, J. J. (1971). Analysis of gas pressure build-up within a porous catalyst particle which is wet by a liquid reactant. *Chem. Engng Sci.* **26**, 533.

KOCHINA, N. N. (1968). Periodic solution of the diffusion equation with nonlinear boundary condition. *Soviet Phys. Dokl.* **13**, 305.

KOLMOGOROFF, A., PETROVSKY, I., and PISCOUNOFF, N. (1937). Etude de l'équation de la diffusion avec croissance de la quantité de matière et son application à un problème biologique. *Bull. Univ. d'Etat Moscou.* **A1**, 6.

KOPELL, N. and HOWARD, L. N. (1973*a*). Horizontal bands in the Belousov reaction.

———, ———. (1973*b*). Spatial structure in the Belousov reaction. External gradients.

———, ———. (1973*c*). Plane wave solutions to reaction-diffusion equations. *Studies in appl. Math.* **52**, 291.

KRASNOSEL'SKII, M. A. (1964). *Topological methods in the theory of nonlinear internal equations.* Pergamon Press, Oxford.

KUIPER, H. J. (1971). On positive solutions of nonlinear elliptic eigenvalue problems. *Rc. Circ. math. Palermo*, Ser II **20**, 113.

LAETSCH, T. W. (1970*a*). Existence and bounds for multiple solutions of nonlinear equations. *SIAM J. appl. Math.* **18**, 389.

———. (1970*b*). The number of solutions of a nonlinear two point boundary value problem. *Indiana Univ. Publ. Sci. Ser., Maths Section* **20**, 1.

LAKSHMIKANTHAM, V. (1964). Parabolic differential equations and Lyapunov-like functions. *J. math. Analysis Applic.* **9**, 234.

LANDAUER, R. (1962). Fluctuations in bistable tunnel diode circuits. *J. appl. Phys.* **33**, 2209.

———. (1971). Stability and instability in information processing and in steady state dissipative systems. *Physik*, 1971. B. G. Tuebner, Stuttgart.

LAPIDUS, L. and BERGER, A. J. (1968). An introduction to the stability of distributed systems via a Liapunov functional. *A.I.Ch.E. Jl.* **14**, 558.

——— and GUREL, O. (1969). A guide to the generation of Liapounov functions. *Ind. Eng. Chem.* **61**(3), 30.

——— and LUUS, R. (1966). An averaging technique for stability analysis. *Chem. Engng Sci.* **21**, 159.

———, ———. (1967). *Optimal control of engineering processes.* Ginn/Blaisdell, Waltham, Massachusetts.

———, PADMANABHAN, L., and LEE, J. C. M. (1972). Stability of an exothermic reaction inside a catalytic slab with external transport limitation. *Ind. Eng. Chem., Fundam.* **11**, 117.

———, ———, and YANG, R. Y. K. (1971). On the analysis of the stability of distributed reaction systems by Lyapounov's direct method. *Chem. Engng Sci.* **26**, 1857.

LEE, T. W. and GOURISHANKAR, V. (1970). On the stability of distributed systems via a Liapounov functional. *Can. J. chem. Engng* **48**, 137.

LIONS, J. L. (1971). *Optimal control of systems governed by partial differential equations.* Springer-Verlag, Heidelberg.

———. (1972). Some aspects of the optimal control of distributed parameter systems. *Regional Conference Series in applied Mathematics* **6**. *SIAM*, Philadelphia, Pennsylvania.

LUSS, D. (1968*a*). Sufficient conditions for uniqueness of the steady state solutions in distributed parameter systems. *Chem. Engng Sci.* **23**, 1249.

———. (1969). On the uniqueness of a large distributed parameter system with chemical reaction and heat and mass diffusion. *Chem. Engng Sci.* **24**, 879.

LUSS, D. (1970). Temperature rise of catalytic supported crystallites. *Chem. Eng. J.* **1**, 311.

———. (1971). Uniqueness criteria for lumped and distributed parameter chemically reacting systems. *Chem. Engng Sci.* **26**, 1713.

———. (1972). Some further observations concerning multiplicity and stability of distributed parameter systems. *Chem. Engng Sci.* **27**, 2299.

——— and CARDOSO, M. A. A. (1969). Stability of catalytic wires. *Chem. Engng Sci.* **24**, 1699.

——— and ERVIN, M. A. (1972*a*). The influence of end effects on the behavior and stability of catalytic wires. *Chem. Engng Sci.* **27**, 315.

———, ———. (1972*b*). Temperature fluctuations (flickering) of catalytic wires and gauzes. Part I. Theoretical investigation. *Chem. Engng Sci.* **27**, 339.

——— and LEE, J. C. M. (1968). On global stability in distributed parameter systems. *Chem. Engng Sci.* **23**, 1237.

———, ———. (1970). The effect of Lewis number on the stability of a catalytic reaction. *A.I.Ch.E. Jl* **16**, 620.

———, ———. (1971). Stability of an isothermal catalytic reaction with complex rate expression. *Chem. Engng Sci.* **26**, 1433.

———, WORLEY, F. L., and EDWARDS, W. M. (1973). Temperature fluctuations (flickering) of catalytic wires and gauzes—II. Experimental study of butane oxidation on platinum wires. *Chem. Engng Sci.* **28**, 1479.

MARTINEZ, H. M. and BAER, R. M. (1973). The algorithm nature of a reaction-diffusion development process. *Bull. Math. Biol.* **35**, 87.

MCGREAVY, C. and SOLIMAN, M. A. The stability of fixed bed catalytic reactors. *Chem. Engng Sci.* **28**, 1401.

——— and THORNTON, J. M. (1970*a*). Generalized criteria for the stability of catalytic reactors. *Can. J. chem. Engng* **48**, 187.

———, ———. (1970*b*). Stability studies of single catalyst particles. *Chem. Eng. J.* **1**, 296.

MCNABB, A. (1961). Comparison and existence theorems for multicomponent diffusion systems. *J. math. Analysis Applic.* **3**, 133.

MEL, H. C. (1964). On the stability of flow-formed inter-faces, and a diffusion-gravity controlled enzyme-substrate reaction. *Chem. Engng Sci.* **19**, 847.

MLAK, W. (1957). Differential inequalities of parabolic type. *Annls pol. math.* **3**, 349.

MURPHY, E. N. and CRANDALL, E. D. (1970). Liapunov stability of parabolic distributed systems: the catalyst particle problem. *J. bas. Engng* **92**, 265.

NARSIMHAN, R. (1954). On the asymptotic stability of solutions of parabolic differential equations. *J. rat. Mech. Analysis* **3**, 303.

NICOLIS, G. (1971). Stability and dissipative structures in open systems far from equilibrium. *Adv. chem. Phys.* **19**, 209.

NISHIMURA, Y. and MATSUBARA, M. (1969). Stability conditions for a class of distributed-parameter systems and their applications to chemical reaction systems. *Chem. Engng Sci.* **24**, 1427.

ORTOLEVA, P. J. and ROSS, J. (1972*a*). Penetration of boundary perturbations in unstable chemical systems. *J. chem. Phys.* **56**, 287.

———, ———. (1972*b*). Response of unstable chemical systems to external perturbations. *J. chem. Phys.* **56**, 293.

———, ———. (1973). Phase waves in oscillatory chemical reactions. *J. chem. Phys.* **58**, 5673.

OTHMER, H. G. (1969). *Interactions of reaction and diffusion in open systems*. Ph.D. Dissertation, University of Minnesota.

PADMANABHAN, L. and VAN DEN BOSCH, B. (1974*a*). Use of orthogonal collocation for the modeling of catalyst particles. I. Analysis of the multiplicity of solutions. *Chem. Engng Sci.* **29**, 1217.

———, ———. (1974*b*). Use of orthogonal collocation methods for the modeling of catalyst particles. II. Analysis of stability. *Chem. Engng Sci.* **29**, 805.

PAO, C. V. (1969). The existence and stability of solutions of nonlinear operator differential equations. *Archs ration Mech. Analysis* **35**, 16.

——— and VOGT, W. G. (1964). On the stability of nonlinear operator differential equations and applications. *Archs ration Mech. Analysis* **35**, 30.

PAO, Y-H. (1964). Unsteady mass transfer with chemical reaction. *Chem. Engng Sci.* **19**, 694.

———. (1965). Unsteady mass transfer with chemical reaction. More general initial and boundary conditions. *Chem. Engng Sci.* **20**, 665.

PERLMUTTER, D. D. (1972). *Stability of chemical reactors*. Prentice-Hall, Englewood Cliffs, New Jersey.

——— and FAHIDY, T. Z. (1966). Dynamics of a surface catalyzed chemical reaction. *Can. J. chem. Engng* **44**, 95.

——— and GURA, I. A. (1965). The stability of nonlinear systems in the region of linear dominance. *A.I.Ch.E. Jl* **11**, 474.

——— and MCGOWIN, C. R. (1971). Regions of asymptotic stability for distributed parameter systems. *Chem. Engng Sci.* **26**, 275.

PETERSEN, E. E. and FRIEDLY, J. (1964). The rate of chemical reaction at the surface of non-porous catalytic sphere in concentration and temperature gradients—II. Stability of nonisothermal solutions. *Chem. Engng Sci.* **19**, 783.

———, ———, and DEVOGELAERE, R. J. (1964). The rate of chemical reaction . . . I. Steady state solutions. *Chem. Engng Sci.* **19**, 683.

PETERSON, L. D. and MAPLE, C. G. (1966). Stability of solutions of nonlinear diffusion problems. *J. math. Analysis Applic.* **14**, 221.

PIS'MEN, L. M. and KHARKATS, Y. I. (1966). Existence and stability of stationary regimes of a chemical reaction on a porous catalyst granule. *Dokl. Akad. Nauk SSSR* **168**, 632. (Translation: *Dokl. (Proc.) Akad. Sci. U.S.S.R., Chemical Technology Section* **168-9**, 88.)

———, ———. (1968*a*). Asymmetric state of a heterogeneous exothermic reaction. *Dokl. Akad. Nauk SSSR* **178**, 901. (Translation: *Dokl. (Proc.) Acad. Sci. U.S.S.R., Chemical Technology Section* **178-9**, 16.)

———, ———. (1968*b*). Stability of steady-state reactions in a porous catalyst particle. *Dokl. Akad. Nauk SSSR* **179**, 397. (Translation: *Dokl. (Proc.) Acad. Sci. U.S.S.R., Chemical Technology Section* **178-9**, 38.)

POORE, A. B. (1972). *Stability and bifurcation phenomena in chemical reactor theory*. Ph.D. Thesis, California Institute of Technology.

POORE, A. B. (1973). Multiplicity, stability and bifurcation of periodic solutions in problems arising from chemical reactor theory. *Archs ration Mech. Analysis* **52**, 358.

———, UPPAL, A., and RAY, W. H. (1974). On the behavior of continuous stirred tank reactors. *Chem. Engng Sci.* **29**, 967.

PRAGER, S. and FRISCH, H. L. (1971). Time lag and fluctuations in diffusion through an inhomogeneous material. *J. chem. Phys.* **54**, 1451.

———, MALONE, G. H. and HUTCHINSON, T. E. (1972). A simple model for diffusion in independent temporally fluctuating pores. *J. theor. Biol.* **36**, 379.

PRATER, C. D. (1958). The temperature produced by heat of reaction in the interior of porous particles. *Chem. Engng Sci.* **8**, 284.

PRIGOGINE, I. and NICOLIS, G. (1967). On symmetry-breaking instabilities in dissipative systems. *J. chem. Phys.* **46**, 3542.

PRODI, G. (1951). Questione di stabilita per equazione non-lineari alle derivate parziali di tipo parabolico. *Atti Accad. naz. Lincei Rc., Ser.* 8 **10**, 365.

PROTTER, M. H. and WEINBERGER, H. (1967). *Maximum principles in differential equations.* Prentice-Hall, Englewood Cliffs, New Jersey.

RABINOWITZ, P. H. (1971). A global theorem for nonlinear eigenvalue problems and applications. *Proc. Symposium on Nonlinear Functional Analysis*, Madison. (Ed. E. H. Zarantonello).

RAY, W. H. (1972*a*). Catalyst particle and fixed bed reactor dynamics—a review. *Proc. 2nd International Symposium on Chemical Reaction Engineering*, Amsterdam.

———. (1972*b*). Non-isothermal catalyst particle performance under selective deactivation. *Chem. Engng Sci.* **27**, 489.

REGIRER, S. A. (1958). The influence of thermal effects on the viscous resistance of a steady uniform flow of liquid. *J. appl. Math. Mech.* **22**, 580.

ROMANOVSKII, Y. M. and SIDAROVA, G. A. (1966). On the influence of diffusion on the attenuation of oscillating chemical reactions, *Oscillatory processes in biological and chemical systems*, a symposium held 21–6 March, 1966. Scientific Publishing House, Moscow. (In Russian.)

ROSEN, G. (1973*a*). Fundamental theoretical aspects of bacterial chemotaxis. *J. theor. Biol.* **41**, 201.

———. (1973*b*). Approximate general solutions to nonlinear reaction-diffusion equations. *Math. Biosci.* **17**, 367.

———. (1973*c*). Travelling periodic waves of chemical activity. *Phys. Lett.* **45A**, 263.

———. (1974). Approximate solution to the generic initial-value problem for non-linear reaction-diffusion equations. *SIAM J. appl. Math.* **26**, 221.

ROSEN, R. (1970). *Dynamical systems theory in biology.* **1**, John Wiley, New York.

ROUGHTON, F. J. W. (1952). Diffusion and chemical reaction velocity in cylindrical and spherical systems of physiological interest. *Proc. R. Soc.* **B140**, 203.

RUCKENSTEIN, E. (1972). Thermodynamic analysis of the stability of supported metal catalysts. *J. Catal.* **26**, 70.

——— and PETTY, C. A. (1972). On the aging of supported metal catalyst due to hot spots. *Chem. Engng Sci.* **27**, 937.

——— and PULVERMACHER, B. (1972). Ageing kinetics of supported metal catalysts. *Proc. 1st Pacific Chem. Engng Congr.* Paper 1–2, Part I, p. 8.

RUELLE, D. (1973). Bifurcations in the presence of a symmetry group. *Archs ration Mech. Analysis* **51**, 136.

SATTERFIELD, C. N., ROBERTS, G. W., and HARTMAN, J. S. (1967). Effects of initial conditions on the steady state activity of catalyst particles. *Ind. Eng. Chem., Fundam.* **6**, 80.

SATTINGER, D. H. (1968). Stability of nonlinear parabolic systems. *J. math. Analysis Applic.* **24**, 241.

———. (1972). Stability of solutions of nonlinear equations. *J. math. Analysis Applic.* **39**, 1.

———. (1971). Stability of bifurcating solutions by Leray-Schauder degree. *Archs ration Mech. Analysis* **43**, 154.

———. (1972). Monotone methods in nonlinear elliptic and parabolic boundary values problems. *Indiana Univ. Publs Sci. Ser., Maths Section* **21**, 979.

———. (1973). *Topics in stability and bifurcation theory*. Lecture notes in mathematics 309. Springer-Verlag, Heidelberg.

———. (1974). A nonlinear parabolic system in the theory of combustion. *Q. appl. Math.* To appear.

SCHMEAL, W. R. and STREET, I. R. (1971). Polymerization in expanding catalyst particles. *A.I.Ch.E. Jl* **17**, 1188.

SCHMITZ, R. A. (1967). A further study of diffusion flame stability. *Combust. Flame* **11**, 49.

——— and GROSBOLL, M. P. (1965). Homogeneous combustor stability. *Combust. Flame* **9**, 339.

——— and KIRKBY, L. L. (1966). An analytical study of the stability of a laminar diffusion flame. *Combust. Flame* **10**, 205.

——— and LINDBERG, R. C. (1969). On the multiplicity of steady states in boundary layer problems with surface reaction. *Chem. Engng Sci.* **24**, 1113.

———, ———. (1970). Multiplicity of states with surface reaction on a blunt object in a convective system. *Chem. Engng Sci.* **25**, 901.

———, ———. (1971). Dynamics of heterogeneous reaction at a stagnation point: numerical study of nonlinear transient effects. *Int. J. Heat Mass Transfer* **14**, 718.

——— and WINEGARDNER, D. K. (1967). Dynamics of heterogeneous reaction at a stagnation point. *A.I.A.A. J.* (*Am. Inst. Aeronaut. Astronaut.*) **5**, 1589.

———, ———. (1968). Stability of reaction on a spherical particle. *A.I.Ch.E. Jl* **14**, 301.

SCRIVEN, L. E., DESIMONE, J. A., and BEIL, D. L. (1973). Ferroin-Collodion membranes: dynamic concentration patterns in planar membranes. *Science* **180**, 946.

——— and GMITRO, J. I. (1966). A physico-chemical basis for pattern and rhythm. *Intracellular Transport* (Ed. K. B. Warren). Symp. Int. Soc. Cell Biol. Academic Press, New York.

——— and OTHMER, H. G. (1969). Interactions of reaction and diffusion in open systems. *Ind. Eng. Chem., Fundam.* **8**, 302.

———, ———. (1971). Instability and dynamic pattern in cellular networks. *J. theor. Biol.* **32**, 507.

———, ———. (1974). Nonlinear aspects of dynamic pattern in cellular networks. *J. theor. Biol.* **43**, 83.

SEGEL, L. A. (1971). On collective motions of chemotactic cells. *Lectures on mathematics in the life sciences.* (Ed. J. Cowan) **4**, p. 3. American Mathematics Society, Providence.

——— and KELLER, E. F. (1970*a*). Conflict between positive and negative feedback as an explanation for the initiation of aggregation in slime mould amoebae. *Nature* **227**, 1363.

———, ———. (1970*b*). Initiation of slime mold aggregation viewed as instability. *J. theor. Biol.* **26**, 339.

———, ———. (1971*a*). Model for chemotaxis. *J. theor. Biol.* **30**, 2.

———, ———. (1971*b*). Traveling bands of chemotactic bacteria: a theoretical analysis. *J. theor. Biol.* **30**, 235.

——— and JACKSON, J. L. (1972). Dissipative structure: an explanation and an ecological example. *J. theor. Biol.* **37**, 554.

——— and STOECKLY, B. (1972). Instability of a layer of chemotactic cells, attractant and degrading enzyme. *J. theor. Biol.* **37**, 561.

SERRIN, J., DOUGLAS, J., and DUPONT, T. (1971). Uniqueness and comparison theorems for nonlinear elliptic equations in divergence form. *Archs ration Mech. Analysis* **42**, 157.

SHENSA, M. J. (1971). Parasitics and the stability of equilibrium points of nonlinear networks. *IEEE Trans. Circuit Theory* **CT-18**, 481.

——— and DESOER, C. A. (1970). Networks with very small and very large parasitics: natural frequencies and stability. *Proc. IEEE* **58**, 1933.

SIMCHEN, A. E. (1964). Stationary temperatures and critical temperatures in exothermal reactions. *Israel J. Tech.* **2**, 248.

SLIN'KO, M. G., BESKOV, V. D., and ZELENYAK, T. I. (1966). Qualitative analysis of equations describing exothermic processes. Catalytic processes in a porous slab. *Kinet. Catal.* **7**, 760.

SMITH, J. M. and MASAMUNE, S. (1966). Performance of fouled catalyst pellets. *A.I.Ch.E. Jl* **12**, 384.

———, ———, and SAGARA, M. (1967). Effect of non-isothermal operation on catalyst fouling. *A.I.Ch.E. Jl* **13**, 1226.

STEWART, W. E. (1968). Diffusion and first-order reaction in time-dependent flows. *Chem. Engng Sci.* **23**, 483.

SZARSKI, J. (1955). Sur la limitation et l'unicité des solutions d'un systeme non-linear d'équations paraboliques aux dérivées partielles du second ordre. *Annls pol. math.* **2**, 237.

———. (1965). *Differential inequalities.* Polish Scientific Publishers, Warsaw.

TAAM, C. T. (1967). On nonlinear diffusion equations. *J. Different Eqat.* **3**, 482.

TATTERSON, D. F. and HUDSON, J. L. (1973). An experimental study of chemical wave propagation. *Chem. Eng. Commun.* **1**, 3.

THOMAS, P. H. (1961). Effect of reactant consumption on the induction period and critical condition for a thermal explosion. *Proc. R. Soc.* **A262**, 192.

——— and BOWES, P. C. (1961). Thermal ignition in a slab with one face at a constant high temperature. *Trans. Faraday Soc.* **57**, 2007.

TURING, A. M. (1952). The chemical basis of morphogenesis. *Phil. Trans. R. Soc.* **B237**, 37.

VILLADSEN, J. and MICHELSEN, M. L. (1972). Diffusion and reaction on spherical catalyst pellets: steady state and local stability analysis. *Chem. Engng Sci.* **27**, 751.

——— and STEWART, W. E. (1969). Graphical calculation of multiple steady states and effectiveness factors for porous catalysts. *A.I.Ch.E. Jl* **15**, 28.

——— and SØRENSEN, J. P. (1969). Solution of parabolic partial differential equations by a double collocation method. *Chem. Engng Sci.* **24**, 1337.

VOLTER, B. V. (1968). The odd number of steady state regimes of chemical reactors. *Theor. Found. chem. Eng.* **2**, 402.

WAKE, G. C. (1969). Uniqueness theorem for a system of parabolic differential equations. *J. Different Eqat.* **6**, 36.

WANG, Y-C. (1973). Review of some mathematical models of non-linear domain dynamics in bulk-effect semiconductors. *J. Inst. Math. its Appl.* **11**, 251.

WARDLAW, C. W. (1953). A commentary on Turing's diffusion-reaction theory of morphogenesis. *New Phytol.* **52**, 40.

———. (1955). Evidence relating to the diffusion-reaction theory of morphogenesis. *New Phytol.* **55**, 39.

WARGA, J. (1962). Relaxed variational problems. *J. math. Analysis Applic.* **4**, 111.

WEI, J. (1965). The stability of a reaction with intra-particle diffusion of mass and heat: the Liapunov methods in a metric function space. *Chem. Engng Sci.* **20**, 729.

———. (1966). On the maximum temperature inside a porous catalyst. *Chem. Engng Sci.* **21**, 1171.

WEISZ, P. B. and HICKS, J. S. (1962). The behaviour of porous catalyst particles in view of internal mass and heat diffusion effects. *Chem. Engng Sci.* **17**, 265.

WESTPHAL, H. (1949). Zur Abschätzung der Lösungen nicht linearer parabolischer Differentialgleichungen. *Math. Z.* **51**, 690.

WEYMANN, H. D. (1967). Finite speed of propagation in heat conduction, diffusion, and viscous shear motion. *Am. J. Phys.* **35**, 488.

WICKE, E. (1961). Stabile und instabile Reaktionszüstande bei exothermen Umsetzungen. *Z. Elektrochem.* **65**, 267.

———, BEUSCH, H. and FIEGUTH, P. (1972). Unstable behavior of chemical reactions at single catalyst particles. *Proc. 1st International Conference on Chemical Reaction Engineering*, Washington 1970, p. 615. American Chemical Society, Washington.

——— and FIEGUTH, P. (1971). Der Übergang vom Zünd/Lösch-Verhalten zu stabilen Reaktionszuständen bei einem adiabatischen Rohrreaktor. *Chemie-Ingr-Tech.* **43**, 604.

——— and HUGO, P. (1968*a*). Transportprozesse and thermische Instabilitäten bei porösen Katalysatoren. *Chemie-Ingr-Tech.* **40** (23), 1133.

———, PADBERG, G. and ARENS, H. (1968). Thermische Instabilitäten bei exothermen katalytischen Gasreaktionen in adiabatischer Kontaktschicht. *Proc. IVth European Symposium on Chemical Reaction Engineering*, Brussels. Pergamon Press, Oxford.

WINEGARDNER, D. K. (1967). *The stability of heterogeneous reactions.* Ph.D. Thesis, University of Illinois.

WINFREE, A. T. (1972). Spiral waves of chemical activity. *Science* **175**, 634.

WOLPERT, L. (1969). Positional information and the spatial pattern of cellular differentiation. *J. theor. Biol.* **25**, 1.

ZEL'DOVICH, Y. B. (1948). Towards a theory of flame propagation. *Z. fiz. Khim.* **22**, 1. (In Russian.)

——— and BARENBLATT, G. I. (1971). Intermediate asymptotics in mathematical physics. *Russ. math. Survs* **26**(2), 45.

——— and FRANK-KAMENETSKII, D. A. (1938). Toward a theory of uniform propagation of a flame. *Dokl. Akad. Nauk SSSR* **19**, 693. (In Russian.)

ZELENYAK, T. I. (1966). On stationary solutions of mixed problems relating to the study of certain chemical processes. *Different Equations* **2**, 98. (Transl. of *Differentsialmye Uravneniya* **2**, 205.)

———. (1967). The stability of solutions of mixed problems for a particular quasilinear equation. *Different Equations* **3**, 9. (Trans. of *Differentsialmye Uravneniya* (1967) **3**, 19.)

ZHABOTINSKII, A. M. (1964*a*). Periodic course of oxidation of malonic acid in solution. Investigation of the kinetics of the reaction of Belousov. *Biofizika* **9**, 306.

———. (1964*b*). Periodic oxidation reactions in the liquid phase. *Dokl. Akad. Nauk SSSR* **157**, 392.

———. (1967). Oscillatory chemical reactions in a homogeneous medium and related problems. In *Oscillatory processes in biological and chemical systems* (Ed. G. M. Frank) Publishing House Nauka, Moscow.

———. (1970). Concentration wave propagation in two-dimensional liquid-phase self-oscillating system. *Nature* **225**, 535.

———, DESHCHEREVSKII, V. I., SEL'KOV, Y. Y., SIDORENKO, N. P., and SHNOL', S. E. (1970). Oscillatory biological processes at the molecular level. *Biofizika* **15**, 235.

———, VAVILIN, V. A., and ZAIKIN, A. N. (1968). Effect of ultraviolet radiation on the oscillating oxidation reaction of malonic acid derivatives. *Russ. J. phys. Chem.* **42**, 1649.

———, ZAIKIN, A. N., KORZUKHIN, M. D., and KREITSER, G. P. (1971). Mathematical model of a self-oscillating chemical reaction oxidation of bromomalonic acid with bromate catalysed by cerium ions. *Kinet. Katal.* **12**, 584.

ZIEN, L. (1973). An upper bound for the singular parameter in a stable, singularly perturbed system. *J. Franklin Inst.* **295**, 373.

ØSTERGAARD, K. (1963). The thermal instability of solid catalysts and the temperature dependence of heterogeneous catalytic process rates. *Chem. Engng Sci.* **18**, 259.

APPENDED ADDITIONAL BIBLIOGRAPHICAL COMMENTS

IN a subject of such lively activity it is inevitable that any book is partially out of date by the time it appears. To overcome some part of this deficiency the publishers have kindly allowed me to add some AABCs to each volume in much the same style as the ABCs to the chapters.

6.2. Some non-existence theorems are given by Levine (1973). For more positive results see Keller and Keener (1974).

6.4.2. For some remarkable results of biological import see Bunow (1974).

6.6. Rabinowitz and Ambrosetti (1973) have some powerful general results. See also Zischka and Chow (1974).

7.1. Some general considerations on bifurcation are given by Crandall and Rabinowitz (1973) and Sattinger (1973). The volume edited by Stackgold, Joseph, and Sattinger (1973) contains several papers of value.

7.4. See also Auchmuty (1973) and Lapidus and Yang (1974). Friedly, Lin, and Kinnen (1973) have used the methods of exterior differential forms to motivate a procedure for generating Liapunov functionals.

7.7. Luss (1974) has considered the effect of capacitance terms, such as the Lewis number, for both discrete and continuous systems and shown that marginal stability is always associated with a pair of imaginary roots.

7.9. See also Ray, Uppal, and Poore (1974).

7.10. See also Aris (1974*b*).

7.11. Some recent work on chemotaxis is given by Nanjundiah (1973), Rosen (1973*a*, 1974), Rubinow (1973), and by Segel and Jackson (1973).

7.12. Glass and Kaufman (1972) considered the interaction of two species which respectively promote and inhibit the formation of the other at the two outside tanks of a chain of four; a stable limit cycle arises. Thames has generalized it to the continuous case and treats this discretely (1973) or by periodic extension with a Green's function (1974). See also Edelstein (1971), Martinez (1972), Jorné (1974), and Rössler (1974) for other models connected with morphogenesis. The most striking model is that of Meinhardt and Gierer (1972, 1974). They consider the diffusion, generation and interaction of two species an activator and an inhibitor and obtain many results that parallel the growth phenomena in hydra.

8.1. An instance of the inversion (8.16) should have been referred to. In the case of the Dirichlet problem for a slab

$$\eta(\phi^2+s,\infty) = \{\tanh(s+\phi^2)^{\frac{1}{2}}\}/\{s+\phi^2\}^{\frac{1}{2}}$$

so that

$$\eta(\tau) = e^{-\phi^2\tau}\theta_2(0|i\pi\tau) = (\pi\tau)^{-\frac{1}{2}} \sum_{n=-\infty}^{\infty} e^{-(\phi^2\tau+n^2/\tau)}.$$

Towler and Rice (1974) have considered the response of spherical catalyst particles immersed in a stirred tank. The three principal parameters are the Thiele modulus, ratio of volumes of particle to stirred tank, and ratio of holding time to the characteristic diffusion time. For certain values of the parameters the concentration can pass through a minimum before levelling off to its steady state.

Linear problems with biological implications are discussed by Jorné (1974) and by Macdonald, Mann, and Sperelakis (1974).

8.2. See also Bidner and Calvelo (1974*a*, *b*).

8.3. Levitsky and Shaffer (1973) have given some approximations to the temperature history for homogeneous exothermic reactions.

8.5. See also Bailey and Lee (1974).

8.8. The same authors (1975) have considered the distributed system in some pilot calculations. They find that the effect of the heat adsorption can be quite dramatic. Thus when the parameters are chosen to give a low temperature steady state, the transient with $\hat{\beta} = 0$ approaches the steady state smoothly. But with $\hat{\beta} = 0{\cdot}05$ a rapid high temperature wave surges to the centre and subsides in a time of the order a^2/D.

8.9. For more recent physically aposite results see Schmidt and Steinbrüchel (1973, 1974).

8.10. Much work has been done in this area during recent months. Field, Koros, and Noyes, (1972*a*, *b*) and Field and Noyes (1974) have worked out a good simplified kinetics of the Belousov reaction. Murray (1974) and Othmer (1975) have explored this model, the so-called Oregonator. The Brusselator has been examined in detail by Boa (1974) and used by Nicholis (1973). See also Nicolis and Auchmuty (1974), Pavlidis (1973), Ross, Ortoleva, and Hahn (1973), Zhabotinskii and Zaikin (1973), and Winfree (1974). Some very nice results have emerged from the work of Kopell and Howard (1974*a*, *b*); the American Mathematical Society Symposium on Applied Mathematics volume in which their 1974*a* paper will appear should have other relevant papers. Keller and Rinzel (1973) have found travelling wave solutions to a nerve conduction equation.

ADDITIONAL REFERENCES

(Incomplete references are to papers known to have been accepted. They will be found in the stated journal for late 1974 or 1975.)

ARIS, R. (1974*b*). Phenomena of multiplicity, stability and symmetry. *Ann. N. Y. Acad. Sci.* **231**, 86.

AUCHMUTY, J. F. G. (1973). Liapounov methods and equations of parabolic type. In Stackgold, Joseph, and Sattinger (1973).

BAILEY, J. E. and LEE, C. K. (1974). Diffusion waves and selectivity modifications in cyclic operation of a porous catalyst. *Chem. Engng Sci.* **29**, 1157.

BIDNER, M. S. and CALVELO, A. (1974). Transient analysis of exothermal reactions within catalyst pellets. Effects of initial conditions. *Chem. Engng Sci.* **29**, 1237.

———, ———. (1974*b*). Effect of Lewis number on the transient behaviour of gas-solid systems. *Chem. Engng Sci.* **29**, 1909.

BOA, J. A. (1974). *A model biochemical reaction*. Ph.D. Dissertation, California Institute of Technology.

BUNOW, B. (1974). Enzyme kinetics in cells. *Bull. Math. Biol.* **36**, 157.

CRANDALL, M. G. and RABINOWITZ, P. H. (1973). Bifurcation, perturbation of simple eigenvalues and linearized stability. *Arch. Rat. Mech. Anal.* **52**, 161.

CRESSWELL, D. L. and ELNASHIE, S. S. E. (1974). The influence of reactant adsorption on the multiplicity and stability of the steady state of a catalyst particle. *Chem. Engng Sci.* **29**, 753.

———, ———. (1975). On the dynamic modelling of porous catalyst pellets: distributed model. *Chem. Engng Sci.* **30**. Not yet published.

EDELSTEIN, B. B. (1971). A cell specific diffusion model of morphogenesis. *J. theor. Biol.* **30**, 515.

FIELD, R. J., KOROS, E., and NOYES, R. M. (1972*a*). Oscillations in chemical systems. I. Detailed mechanism of a system showing temporal oscillations. *J. Am. chem. Soc.* **94**, 1394.

———, ———, ———. (1972*b*). Oscillations in chemical systems. II. Thorough analysis of temporal oscillations in the bromate-cerium-malonic acid system. *J. Am. chem. Soc.* **94**, 8649.

——— and NOYES, R. M. (1974). Oscillations in chemical systems. IV. Limit cycle behaviour in a model of a real chemical reaction. *J. chem. Phys.* **60**, 1877.

FRIEDLY, J. C., LIN, Y. H., and KINNEN, E. (1973). Construction of Liapounov functionals for partial differential equations using exterior differential forms. *Proc. J.A.C.C.*, p. 864, Columbus, Ohio.

GLASS, L. and KAUFFMAN, S. A. (1972). Co-operative components, spatial localization and oscillatory cellular dynamics. *J. theor. Biol.* **34**, 219.

JORNÉ, J. (1974). The effects of ionic migration on oscillations and pattern formation in chemical systems. *J. theor. Biol.* **43**, 375–380.

KELLER, H. B. and KEENER, J. P. (1974). Positive solutions of convex nonlinear eigenvalue problems. *J. diff. Eqns.* **16**, 103.

KELLER, J. B. and RINZEL, J. (1973). Travelling wave solutions of a nerve conduction equation. *Biophys. J.* **13**, 1313.

KOPELL, N. and HOWARD, L. N. (1974*b*). Pattern formation in the Belousov reaction. *Proc. Symp. Some mathematical questions in biology* AMS-SIAM. San Francisco.

———, ———. (1974*a*). Wave trains, shock structures and transition layers in reaction-diffusion equations. *Proc. Symp. App. Math.* American Mathematical Society. Not yet published.

LEVINE, H. A. (1973). Some nonexistence and instability theorems for solutions of formally parabolic equations of the form $Pu_t = -Au + F(u)$. *Arch. Rat. Mech. Anal.* **51**, 371.

LAPIDUS, L. and YANG, R. Y. K. (1974). Lyapounov stability analysis of porous catalyst particle systems: extension to interphase transfer and time-varying surface conditions. *Chem. Engng Sci.* **29**, 1567.

LEVITSKY, M. and SHAFFER, B. W. (1973). The approximation of temperature distributions in homogeneous exothermic reactions. *Chem. Eng. J.* **5**, 235–42.

LUSS, D. (1974). The influence of capacitance terms on the stability of lumped and distributed parameter systems. *Chem. Engng Sci.* **29**, 1832.

MACDONALD, R. L., MANN, J. E., and SPERELAKIS, N. (1974). Derivation of general equations describing tracer diffusion in any two-compartment tissue with application to ionic diffusion in cylindrical muscle bundles. *J. theor. Biol.* **45**, 107.

MARTINEZ, H. M. (1972). Morphogenesis and chemical dissipative structures. *J. theor. Biol.* **36**, 479.

MEINHARDT, H. and GIERER, A. (1972). A theory of biological pattern formation. *Kybernetik* **12**, 30–9.

———, ———. (1974). Applications of a theory of biological pattern formation based on lateral inhibition. *J. Cell. Sci.* **15**. Not yet published.

MURRAY, J. D. and HASTINGS, S. P. (1974). The existence of oscillatory solutions in the Field–Noyes model of the Belousov–Zhabotinski reaction. *SIAM J. Appl. Math.* **26**. Not yet published.

NANJUNDIAH, B. (1973). Chemotaxis, signal relaying and aggregation morphology. *J. theor. Biol.* **42**, 63.

NICOLIS, G. (1973). Mathematical problems in theoretical biology. In Stackgold, Joseph, and Sattinger (1973).

———, and AUCHMUTY, J. F. G. (1974). Dissipative structures, catastrophes and pattern formation: a bifurcation analysis. *Proc. Natn. Acad. Sci.* Not yet published.

OTHMER, H. G. (1975). On the temporal characteristics of a model for the Zhabotinskii–Belousov reaction. *Math. Biosci.* Not yet published.

PAVLIDIS, T. (1973). *Biological oscillators: their mathematical analysis.* Academic Press, New York.

RABINOWITZ, P. and AMBROSETTI, A. (1973). Dual variational methods in critical point theory and its applications. *J. funct. Anal.* **14**, 349.

RAY, W. H., UPPAL, A., and POORE, A. B. (1974). On the dynamic behaviour of catalytic wires. *Chem. Engng Sci.* **29**, 1330.

ROSEN, G. (1974a). On the propagation theory of bands of chemotactic bacteria. *Math. Biosci.* **20**, 185.

———, (1974b). Necessary conditions for the existence of periodic solutions to systems of reaction–diffusion equations. *Math. Biosci.* **21**, 345.

ROSS, J., ORTOLEVA, P. J., and HAHN, S-S. (1973). Chemical oscillations and multiple steady states due to variable boundary permeability. *J. theor. Biol.* **41**, 503.

RÖSSLER, O. E. (1974). Chemical automata in homogeneous and reaction-diffusion kinetics. *Proc. Internat. Summer School in the Physics and Mathematics of the Nervous System*, Springer-Verlag, Heidelberg.

RUBINOW, S. I. (1973). Mathematical problems in the biological sciences. *Regional conference series in applied mathematics*, Vol. **10**. SIAM, Philadelphia.

SATTINGER, D. H. (1973). Six lectures on the transition to instability. In Stackgold, Joseph, and Sattinger (1973).

SCHMIDT, L. D. and STEINBRÜCHEL, Ch. (1973). Heat dissipation in catalytic reactions on supported catalysts. *Surf. Sci.* **40**, 693.

———, ———. (1974). Energy dissipation in catalysis. *J. Vac. Sci. Technol.* **11**, 267.

SEGEL, L. A. and JACKSON, J. L. (1973). Theoretical analysis of chemotactic movement in bacteria. *J. Mechanochem. Cell Motility* **2**, 25–34.

STACKGOLD, I., JOSEPH, D. D., and SATTINGER, D. H. (1973). Nonlinear problems in the physical science and biology. *Lecture Notes in Maths*, Vol. 322. Springer-Verlag, Heidelberg.

THAMES, H. D. (1973). Dependence of stability on inter-enzyme distance in cooperative systems. *J. theor. Biol.* **41**, 331.

———. (1974). Stability and enzyme separation; integral representation of the solutions. *Bull. Math. Biol.* **36**, 197.

THOM, R. (1968). Une théorie dynamique de la morphogenèse. In *Towards a theoretical biology*, I Prolegomena (ed. C. H. Waddington), p. 152. Edinburgh University Press, Edinburgh.

TOWLER, B. F. and RICE, R. G. (1974). A note on the response of a CSTR to a spherical catalyst pellet. *Chem. Engng Sci.* **29**, 1828.

WEI, J. and ADAM, D. E. (1974). Mass Transport of ATP within the motile sperm. *J. theor. Biol.* Not yet published.

WETTE, R., KATZ, I. N. and RODIN, E. Y. (1974). Stochastic processes for solid tumor. Diffusion regulated growth. *Math. Biosci.* **21**, 311.

WHEELER, J. M. and MIDDLEMAN, S. (1970). Machine computation of transients in fixed beds with intraparticle diffusion and nonlinear kinetics. *Ind. Eng. Chem. Fundam.* **9**, 624.

WINFREE, A. T. (1974). Rotating chemical reactions. *Scient. Am.* **230**, 82–95.

ZHABOTINSY, A. M. and ZAIKIN, A. N. (1973). Autowave processes in a distributed chemical system. *J. theor. Biol.* **40**, 45–61.

ZISCHKA, K. A. and CHOW, P. S. (1974). On nonlinear initial boundary value problems of heat conduction and diffusion. *SIAM Review* **16**, 17.

AUTHOR INDEX

SUBJECT INDEX